Editor
Kim Fields

Illustrator
Kelly McMahon

Cover Artist
Brenda DiAntonis

Editorial Project Manager
Mara Ellen Guckian

Managing Editor
Ina Massler Levin, M.A.

Creative Director
Karen J. Goldfluss, M.S. Ed.

Art Production Manager
Kevin Barnes

Art Coordinator
Renée Christine Yates

Imaging
James Edward Grace
Craig Gunnell
Rosa C. See

Publisher

Mary D. Smith, M.S. Ed.

W9-BTL-934

Author

Laurie Hansen, M.S. Ed.

Teacher Created Resources, Inc.
6421 Industry Way
Westminster, CA 92683
www.teachercreated.com
ISBN: 978-1-4206-8771-2
© 2007 Teacher Created Resources, Inc.
Made in U.S.A.

Table of Contents

Table of Contents *(cont.)*

Introduction

Science Through the Year: Grades 1–2 was designed for busy teachers who would like to integrate simple, inquiry-based science units into their curriculum. All of the activities included in this book will help students develop positive attitudes toward science and content learning in the areas of physical science, earth science, and life science and health. Text for students is written on a listening-comprehension level.

Science experiences in the early elementary grades lay the foundation for scientific literacy in upper elementary, middle school, and beyond. Rich science experiences help children develop positive scientific attitudes such as curiosity, enthusiasm, and open-mindedness. It is critical that young children learn science through inquiry, which means that they participate in hands-on learning experiences that allow them to develop science process skills such as observing, classifying, measuring, predicting, comparing/contrasting, and planning an investigation.

Children will develop sensory and teamwork skills as they complete interdisciplinary activities in *Science Through the Year: Grades 1–2*. Students will become excited about science as they learn key concepts and vocabulary through firsthand experience. They will utilize the same skills scientists use every day. The inquiry activities in this book address the auditory, kinesthetic, and visual modalities and are appropriate for English learners and students with special needs. These skills, along with the motivation the children will experience, will help them get an inside look at the wonderful world of science.

Introduction *(cont.)*

Each unit lists the applicable National Science Education Standards, as well as grade-specific content standards, and includes the following components:

Picture Charts—one or more blackline illustrations to provide visuals for concepts being presented.

Teacher Notes—offers new science vocabulary and content information to be used in discussions and with the Picture Charts. The scientific vocabulary and definitions should be adjusted to meet student needs.

Home/School Connections—includes worksheets and parent letters. The letters can be sent home at the beginning of the unit, describing the topic being presented and materials needed. The letters offer guides for parent/child discussions of the topic. Worksheets can be used to reinforce skills practiced during class time.

Unit Preparation—lists materials required for the unit and a brief explanation of preparing the unit.

Background Information—provides a brief summary of the content to be presented in teacher-friendly language.

Literature Links—lists related fictional and nonfiction books.

Inquiry-based Activities—provides enough activities for 2–4 weeks of instruction, depending on how often the teacher presents science during the week. Students use inquiry skills such as observation, measurement, prediction, comparing/contrasting, and planning an investigation.

Word Cards—specific to each unit, they can be used for labeling, vocabulary practice, and in writing centers. These cards are especially helpful to English language learners.

Journal Pages—focuses on one or more inquiry skills and provides children a way to express their learning on paper, at the same time providing an ongoing assessment for the teacher.

Minibooks—provides an opportunity for the teacher to integrate content and language. These books also provide opportunities for review and additional home-school connections when children take them home and "read" them with family members.

Assessments—each unit has appropriate assessments embedded within. Teachers can use these to discover what the children already know, gauge how well the children are acquiring vocabulary and concepts, and decide when to move on or reteach content. The included Culminating Assessments provide the teacher with a way to determine what the children have learned over the course of a unit.

How to Use This Book

Use *Science Through the Year: Grades 1–2* to meet your standards, organize class sessions, and coordinate home/school connection activities. Before you begin each unit, you will need to gather materials. Most of the materials are inexpensive and easy to find. Some of the materials, such as ladybird beetle larvae, must be ordered in advance. Be prepared!

Prior to starting each unit, you will need to prepare each of the following:

1. **Science Journals**—Reproduce, collate, add a construction-paper cover, and staple a Science Journal for each child.

2. **Word Cards**—Reproduce the cards on cardstock and/or laminate.

3. **Materials for Hands-on Activities**—Grocery, discount, and pet stores are good sources for materials. Online companies provide access to live insect larvae. Specific suggestions are provided in the units for items that need to be specially ordered. Other basic materials, such as construction paper, glue, and scissors, should be readily available in the classroom.

4. **Minibooks**—Reproduce and assemble a minibook for each student.

5. **Charts**—Gather chart paper and colored markers to prepare the charts used throughout each unit.

6. **Literature Links**—Visit your local library to gather books related to the unit. See Literature Links in each unit for ideas.

7. **Picture Charts**—Create the picture charts prior to the beginning of each unit.

Cooperative Learning Groups

For cooperative learning activities, students should work in small groups of four.

1. Assign jobs to members of each group: Recorder, Materials Manager, Reporter, and Leader.

   - The Recorder writes important information on the chart paper.
   - The Materials Manager is the only person in the group who is allowed to gather materials.
   - The Reporter presents what the group did/learned to the whole class at the end of the lesson.
   - The Leader keeps the group on task, makes sure all group members are participating, keeps track of the time, and raises his or her hand for help if a problem or question arises.
   - All members of each group participate by drawing on the chart paper, discussing concepts, and observing.

2. Groups work together to complete a task on corresponding chart paper or a Science Journal page.

How to Use This Book *(cont.)*

Picture Charts

Picture charts provide children with a visual as they learn vocabulary and content. They also allow teachers to model drawing skills. The illustrations are simple; and, by tracing them, a teacher can describe what he or she is doing, helping students realize that they too can draw, observe, and note data. These diagrams lend themselves well to single concepts (e.g., ladybird beetle), parts compared to a whole (e.g., parts of a plant), or the depiction of a science cycle (e.g., life cycle of an insect).

Preparing a Picture Chart

1. Gather chart paper, pencil, and markers.

2. Enlarge the blackline master.

3. Lay a sheet of butcher paper or chart paper atop the enlarged blackline drawing and lightly trace it using a pencil. (During group time, this illustration will again be traced over using a felt pen.)

4. Lightly write "chunks" of information about the concept, including key concepts and vocabulary, near the appropriate parts of the drawing using a pencil (see Teacher Notes in each unit).

Presenting a Picture Chart

1. Gather the children on the rug or at their desks where they can see the "blank" chart paper.

2. Talk about the concept, diagram, or cycle you have prepared. As you discuss chunks of information, trace over your drawing with a permanent marker.

3. Ask the children to tell you what they just learned about the concept or cycle you just presented. To create a low-risk environment, it is a good idea to request volunteers. Write the children's responses in permanent marker next to the corresponding section of the drawing. Use your own pencil notes to remind yourself of any key points or vocabulary the children missed. Continue until the children have restated all the key points and vocabulary.

4. Post the completed Picture Chart on a classroom wall for future reference.

Standards

Unifying Concepts and Processes

The following concepts relate to primary grade science:

- *Systems and Interactions*—Students study various systems (e.g., an organism) and learn how these systems interact with each other.
- *Continuity and Change*—Students study changes that take place in a plant or animal's life cycle (e.g., metamorphosis), comparing and contrasting with things that stay the same (e.g., number of legs an insect has).
- *Evidence and Explanation*—Students collect evidence through observation and use it to make predictions and inferences.
- *Form and Function*—Students understand form by studying objects, plants, and insects. They recognize the properties of objects and that the parts of plants and insects serve specific purposes.

Science as Inquiry

Definition: the multifaceted process of studying the natural world and solving problems through investigation and experimentation

Specific Standards

- Students understand and use inquiry skills (e.g., observe, classify, describe, measure, predict, compare, contrast, plan).
- Students understand scientific concepts by participating in inquiry-based activities.

Life Science

Definition: the study of living organisms and life processes

Specific Standards

- Students understand the characteristics of organisms.
  - What plants and animals are
  - Parts of plants and animals
- Students understand the life cycles of organisms.
  - Plants
  - Insects
- Students understand the relationship between organisms and their environments.
  - What plants and animals need to survive

Standards *(cont.)*

Physical Science

Definition: the study of matter and energy and their interactions

Specific Standards

- Students understand the properties of matter.
- Forms (states) of matter: liquid, solid, gas
- Properties of liquids, solids, gases
- Properties can change when matter is heated, cooled, or mixed

Earth Science

Definition: the study of the earth and its processes

Specific Standards

- Students understand how changes in the earth and sky affect them.
- The sun warms the earth, sky, and water
- Tools to measure the weather
- Seasons are predictable and weather changes from day to day

Health Science

Definition: the study of the body, disease, and personal health

Specific Standards

- Students accept personal responsibility for personal health.
- Ways in which personal health can be enhanced and maintained
- Behaviors that can prevent disease

Based on National Science Education Standards

Parts of a Seed

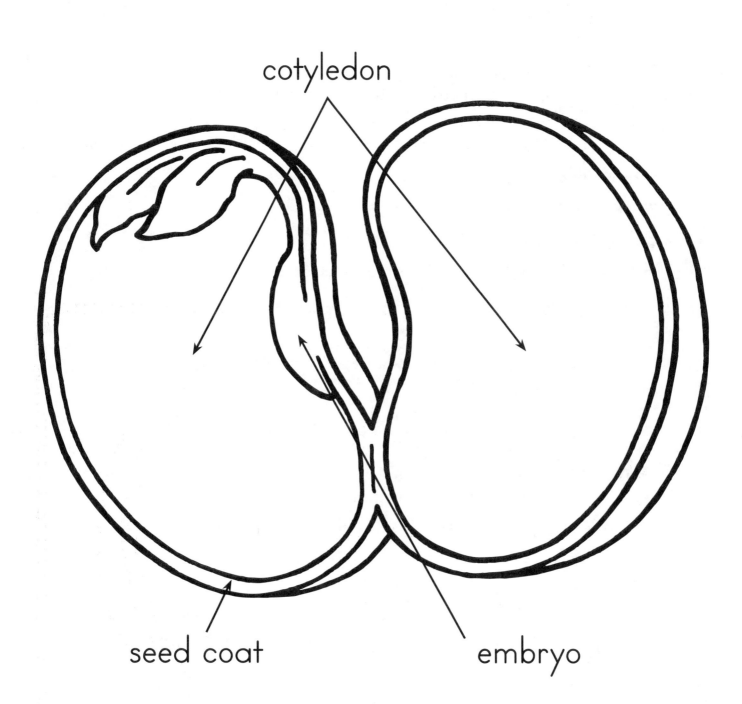

cotyledon

seed coat   embryo

Parts of a Seed Vocabulary

Cotyledon—two thick halves of the lima bean seed

- provides food for the embryo
- becomes the plant's first pair of leaves
- softens once the seed gets water

Embryo—baby plant

- contains all the parts needed to become a new plant
- grows to become the roots and stem of the plant
- grows when water is added and with food provided by the cotyledon

Seed Coat—outer covering of a seed

- protects the seed
- comes off once the seed gets water and the plant begins to grow

Germinate—when a seed sprouts and begins to grow

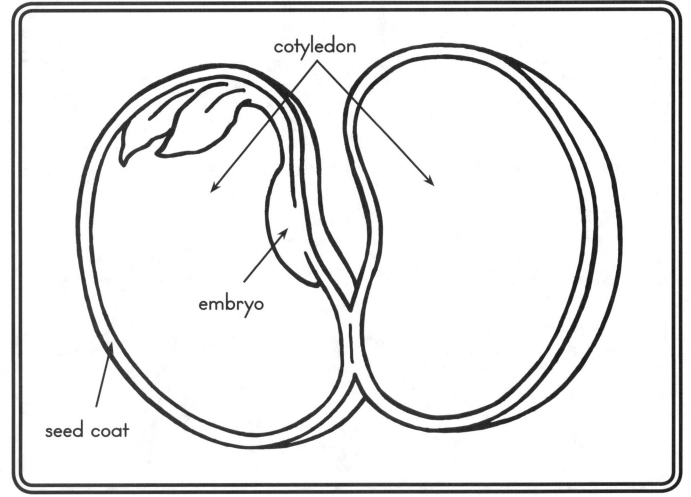

Parts of a Plant

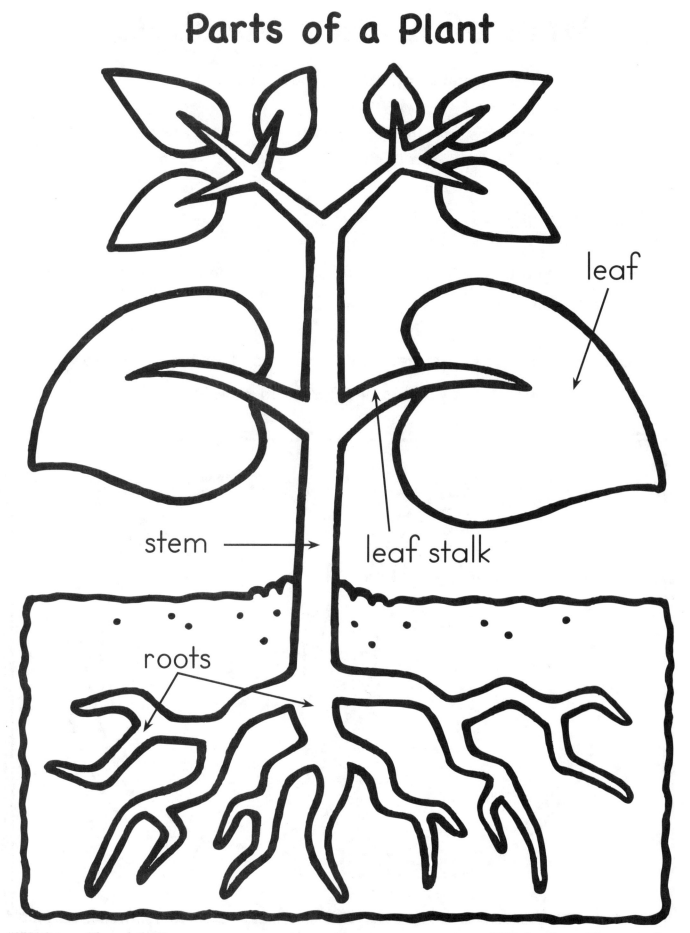

leaf

stem

leaf stalk

roots

Parts of a Plant Vocabulary

Leaf—flat, oval, or round structure that grows on a leaf stalk
- collects sunlight and carbon dioxide
- uses sunlight, water, and carbon dioxide to make food (sugar) for the plant
- is green because it contains chlorophyll

Leaf Stalk—little stem, connecting the leaf to the plant stem

Stem—thin long stalk, main part of a plant
- leaves grow out from the stem through the leaf stalk
- carries water and food from the roots to the leaves and flowers

Root—grows down into the soil
- anchors a plant to the soil
- absorbs water and nutrients from the soil

Shoot—new growth of a plant
- becomes the stem if it comes from the seed
- becomes a leaf stalk if it comes from the stem

Seed Plant—the baby plant that grows from the seed

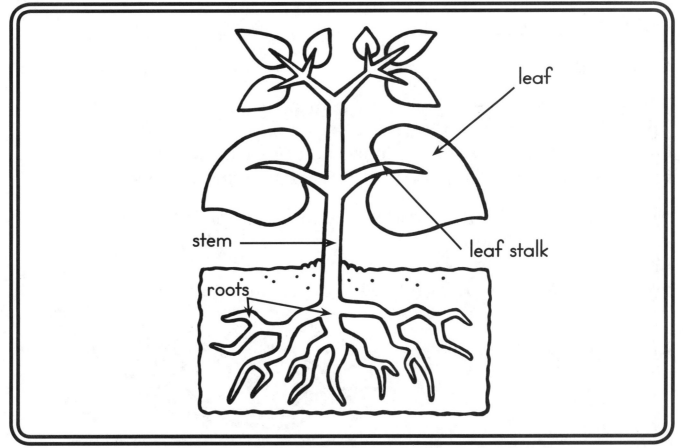

- -

Date _____

Dear Parent,

We will be starting a unit on plants soon. There are a few items we need for the unit. Please provide the item circled below. If you are unable to provide the item, please let me know as soon as possible.

- roll of paper towels
- box of sandwich-sized resealable, plastic bags
- small bag of dried lima beans
- packet of flower seeds
- box of toothpicks
- small Styrofoam cups
- small bag of potting soil

Please send your item to school with your child by _____.
Thank you for your help with this unit!

Sincerely,

- -

Date _____

Dear Parent,

We have just started a science unit on plants! We will grow lima bean plants from seeds to learn what plants need to grow. Over the next few weeks, please ask your child to tell you about what plants need to grow and how plants grow from seeds.

At the end of the unit, students will have the option to take home their lima bean plants, as well as the *Seeds* and *Plants* Minibooks they have been reading in class. Please ask your child to "read" his or her books to you and tell you about his or her plant.

Sincerely,

Plants

 Background Information

Plants are alive. Plants have predictable life cycles. Many plants begin their life cycles in the form of seeds, grow to become seed plants (seedlings), and then adult plants. The last stage is when the plant flowers and produces new seeds. Both seeds and plants have specific structures necessary for growth such as a seed coat (thin protective covering), roots, and leaves.

Students can observe changes over time in seeds and the plants that sprout from them. Students will discover that plants need air, water, and sunlight to survive. Also, students will use inquiry skills such as observation, measurement, prediction, comparing/contrasting, and planning an investigation, when studying plants.

 Unit Preparation

1. Purchase the following items or ask for donations (see letter on page 14):
   - toothpicks
   - hand lens
   - roll of paper towels
   - bag of large dried lima beans
   - box of small, resealable plastic bags (sandwich-size)
   - Styrofoam cups
   - potting soil
   - spray bottles

2. Copy and cut out the *Seeds* and *Plants* Word Cards (pages 40–42).
3. Copy and assemble the *Seeds* and *Plants* Minibooks (pages 34–39).
4. Copy and assemble the *Plants* Science Journals (pages 25–33).
5. Reproduce the *Plants* Culminating Activity and Assessment sheets (pages 23–24).
6. Use chart paper or butcher paper to create the Inquiry Chart for Lesson 1 (see page 16), as well as the whole-class charts used throughout the unit.
7. Make a sample Lima Bean Bag (see page 18) to show the class. Fold a paper towel in half twice and insert into a small, resealable plastic bag. Place two dried lima beans in the bag between the bag and paper towel. Seal the bag.

Literature Links

Anno's Magic Seeds by Mitsumaso Anno
From Seed to Plant by Gail Gibbons
Growing Vegetable Soup by Lois Ehlert
How Do Apples Grow? by Betsy Maestro
Jack and the Beanstalk (Traditional)
Planting a Rainbow by Lois Ehlert
The Carrot Seed by Ruth Krauss
The Magic School Bus Plants Seeds by Joanna Cole
The Tiny Seed by Eric Carle

Plants *(cont.)*

Lesson 1: Introducing the Plant Unit

Ask the children if they have ever grown plants at home. Ask them to turn to a neighbor and tell him or her about the plants the student has grown at home. Allow a few students to share their examples with the whole class. Tell them that today we are starting a science unit about plants, but first, we need to find out what we already know. Show the class the Plants Inquiry Chart (see below) and ask them to help you complete each section of the chart. (Whole Class Assessment)

Note: Color-code each column of the chart using bright markers such as red, blue, and green.

Plants Inquiry Chart

What Do You Know About Plants?	Examples of Plants We Know	What Do You Want to Learn About Plants?	How Can We Find Out?

The inquiry chart serves as an assessment. It demonstrates what the children already know, what their misconceptions may be, and what they are wondering about. It is an excellent springboard for future lessons. Inquiry charts can provide teachers with needed information when planning lessons.

The Sample Plants Inquiry Chart (below) gives an idea of the kinds of responses first grade children might give.

Sample Plants Inquiry Chart

What Do You Know About Plants?	Examples of Plants We Know	What Do You Want to Learn About Plants?	How Can We Find Out?
Plants have flowers.	Apple tree	How do they grow?	Read books.
Some plants have fruit.	Flowers	Is a tree a plant?	Go to the library.
Plants are green.	Roses	How do trees breathe carbon dioxide?	Check the Internet.
You need water and a seed and dirt.		What is carbon dioxide?	Ask our parents.
		I want to learn about planting seeds.	Ask the teacher.
			Grow a plant!

Note: At the end of each unit, you can have the children help you complete a chart of "What We Learned" as a whole group culminating assessment.

Plants *(cont.)*

Lesson 2: Dissecting Seeds

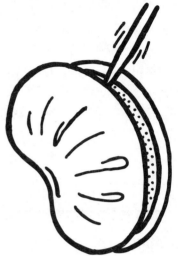

1. Soak 20–30 lima beans overnight. The seeds will soften and expand. When soaked, the parts of the seed become apparent.

2. Distribute two lima beans, a toothpick, and a hand lens to each child.

3. At their desks or tables, the children will dissect their lima bean seeds by carefully squeezing and peeling away the outer layer and then splitting the seeds in half using a toothpick.

4. The children can use the hand lenses to observe the internal structures of the seeds. Ask the children to describe the parts they see. Record student responses on the board or on a sheet of blank chart paper.

5. Have the children draw a picture of the dissected seed in their *Plants* Science Journal and label the parts using their own words (Observation 1, page 26).

Inquiry Skill—Observation

Lesson 3: *Parts of a Seed* Picture Chart

1. Gather the children together on the rug near the *Parts of a Seed* Picture Chart (see page 10) to teach content and vocabulary. Show the children the chart paper with the pre-drawn pencil diagram of the parts of a seed. (Be sure to write your notes lightly in pencil next to each part of the diagram.)

2. Talk about the parts of a seed in "chunks," providing key vocabulary (*seed coat, cotyledons, embryo*) and content (*the seed coat protects, the two cotyledons provide food, the embryo is the baby plant*). As you speak, trace over the corresponding parts with a permanent marker. Do not label the diagram yet.

   Teacher Note: Describe what you are doing as you draw. Mention the curving lines, larger and smaller shapes, etc. Your descriptions should help students feel more comfortable with their own Science Journal illustrations.

3. After tracing over the entire picture and providing the vocabulary and content, ask the children to tell what they learned about the parts of a seed.

4. Write the children's responses on appropriate places on the chart (i.e., labeling parts).

5. Introduce the *Parts of a Seed* Word Cards (page 40) and add them to a pocket chart.

 ***Parts of a Seed* Home Link:** Have the children complete the *Parts of a Seed* worksheet (page 43) for homework.

Plants *(cont.)*

Lesson 4: Lima Bean Bags

Review the Plants Inquiry Chart (see page 16) from Lesson 1. Ask the children if they know what plants need to live and grow. Record their responses on chart paper using a simple web.

Tell the children that they are each going to get two dry seeds to discover what the seeds need to grow. They will observe (watch) the seeds each day over the next few weeks. Some days, they will write their observations in their own *Plants* Science Journals.

Materials (for each child)

- paper towel
- 2 dried lima beans
- small, resealable plastic bag
- 1 tablespoon water
- access to spray bottle

Directions

1. Show the children a sample Lima Bean Bag (see page 15). Explain how to fold the paper towel in half twice, insert it into the bag, and drop the seeds in so that they are between the front of the bag and the paper towel.

2. Distribute two lima bean seeds, a bag, and a half sheet of paper towel to each child. Have the children assemble their lima bean bags.

3. Using a permanent marker, write each child's name on his or her bag.

4. Allow each child to water his or her seeds using one tablespoon of water.

5. Instruct each child to draw a picture of the contents of their lima bean bag in their *Plants* Science Journal and write two sentences predicting what they think their bean seeds will look like tomorrow (Observation 2, Prediction, page 27).

6. The teacher hangs the bags on a window that gets sunlight. The bags should remain sligthly open for air circulation.

7. Each day, have the children check their bags and water as necessary using a spray bottle.

Inquiry Skills—Observation, Predicting

Teacher Tip: It should take about one week for the lima bean seeds to sprout and another week or two for the shoot to grow to include leaves and for the roots to grow. Once the seed turns green and the leaves emerge, it is time to plant the seeds in soil.

Make sure your lima beans get enough sunlight and water—but not too much. Spray the seeds with water if the towel is completely dry. Indirect sunlight or partial day, direct sunlight works better than all-day direct sunlight. If it looks like the beans are getting moldy or are too wet, change the paper towel and let the beans dry a bit before adding more water. The beans will initially sprout without sunlight, but they won't continue to grow without it.

Plants *(cont.)*

Lesson 5: Measuring Seeds

Take the children to the window to observe their seeds. Ask, "What has happened?" Overnight, the seeds should have soaked up the water and expanded. Compare a dry seed to the wet seeds. Ask the children to describe the difference between the two seeds. Record their responses on chart paper or on the board.

Distribute a dry seed and Lima Bean Bag to each child. Demonstrate how to measure the length of dry and wet seeds using a metric ruler. Have the children record their measurements and describe what happened to their lima bean seeds overnight in their *Plants* Science Journals (Observation 3, page 28).

Inquiry Skills—Measurement, Observation, Recording Data

Lesson 6: Literature Connection

1. Introduce a version of the story *Jack and the Beanstalk* to the whole class. Talk about the front cover of the book, the author, and illustrator.

2. Do a "picture walk" through the book. Show the children each page of the book without reading the text. Ask them to predict what will happen at various points in the story.

3. Read the story to the class.

4. Ask the children to share what they liked best about the story.

5. Have the children help you complete a Plot Analysis Chart (see below).

Plot Analysis Chart

Beginning	Middle	End

6. Draw the children's attention to the importance of the beanstalk in the story. Point out the incredible growth of the stalk and ask the children if this could happen in real life. Have the children tell you about the beanstalk's stages of growth (*seed, shoot, large plant, died—when chopped down by Jack*). Have the children take turns retelling the story and acting it out.

7. If you are using this for language arts, you can do a Story Elements Chart (see below). (Whole Class Assessment)

Story Elements Chart

Setting	Characters	Character's Personality Traits

Extension: Once the children have observed the growth of their own seed plants, ask them to write a fictional story about what might happen if their plant had grown like the beanstalk in the story. Each story could be a twist on the original using the child's name (*e.g., Patrick and the Beanstalk*). Children can go through the writing process of brainstorming, rough draft, editing, revising, final draft, and illustrating. (Assessment)

Plants *(cont.)*

Lesson 7: Observing Seeds

Part 1: The children should observe their seeds each day. After a few days, the seeds will sprout. Once they do, discuss the changes with the children and have them observe any changes in their lima bean seeds and record these in their *Plants* Science Journal (Observation 4, page 29).

Inquiry Skill—Observation

Part 2: Have the children look at the dry seeds again. What is the difference so far between the dry seeds and the seeds in their bags? (*These sprouted while the dry seeds did not.*) Ask the children why they think their seeds are sprouting, but the dry seeds are not. Record their responses on a class chart or the board. Have the children compare and contrast the dry seeds and the seeds in their bags in their *Plant* Science Journals (Observation 5, page 30). Have them predict what they think will happen to their seeds in the coming weeks.

Inquiry Skills—Observation, Comparing and Contrasting, Predicting

Lesson 8: *Seeds* Minibook and Big Book

1. Distribute a *Seeds* Minibook (pages 34–36) to each child.

2. Read the book as a class. Discuss the illustrations. Point out key vocabulary (*cotyledon, embryo, seed coat*). Review the *Parts of a Seed* Word Cards (page 40).

3. Explain what the children are to do at each center.

   Center 1: Read the *Seeds* Minibook in small groups with the teacher. The children can take turns reading their Seeds Minibook to each other.

   Center 2: Color the pictures in the *Seeds* Minibook.

   Center 3: Use a yellow crayon to highlight key vocabulary: *cotyledon, embryo, seed coat.* Using a red crayon, circle the high frequency word *the.* Draw a box around the word *seed* each time you see it. (Assessment)

   Center 4: Listen to a taped version of *Jack and the Beanstalk* or the *Seeds* Minibook.

4. At the end of the unit, each child will take home the *Seeds* Minibook and "read" it to his or her family.

5. Create an additional class book by enlarging the *Seeds* Minibook. Students can take turns coloring the pages. Place the completed *Seeds* Big Book in the classroom library.

Lesson 9: Observing the Seed Plants

Have the children observe any changes in their lima bean seeds over the next week or two and record these changes in their *Plants* Science Journals (Observations 6–7, page 31). Once there is a significant change in the sprouts (e.g., the cotyledon turns green), have students draw pictures of how their seed plants look in their *Plants* Science Journals (Observation 8, page 31). Students can also measure their seed plants on two different days and compare the growth.

Inquiry Skills—Observation, Measuring, Comparing and Contrasting

Plants *(cont.)*

Lesson 10: **Planting Seed Plants**

Once the cotyledon turns green, and the plant has developed roots and a stem, it is time to plant the lima bean plant in soil.

1. Provide each child with a Styrofoam cup (with drainage holes punched in the bottom) and his or her Lima Bean Bag. Have them write their names on the cups using markers.

2. Call the children up one at a time or in small groups to fill the cup three-quarters full with potting soil. Children use their fingers to press holes in the soil. Have the children carefully plant their lima bean plants in the hole and cover the roots with more soil.

3. Invite the students to take the plants home and care for them.

Lesson 11: **What Do Plants Need to Survive?**

Now that children have observed their lima bean seeds growing and changing over time, discuss what plants need to survive.

1. Brainstorm with the whole class. Ask, "What did our plants need in order to grow?"

2. Record student responses on chart paper using words and pictures.

3. Facilitate this discussion when the children get stuck. For example, you could ask, *Where did we hang the bags?* (on a sunny window); *What did we give to the seed almost every day?* (water); *Why did we keep the bag unsealed?* (to give the seed air); and *Why did we plant the seeds in soil?* (so the plant could get food [nutrients] from the soil). Record additional ideas on the chart paper.

4. The chart should include words and pictures, illustrating the scientific concept that plants need air, water, food, and sunlight to grow and survive.

Inquiry Skill—Observation

Lesson 12: *Parts of a Plant* **Picture Chart**

1. Gather the children together on the rug near the *Parts of a Plant* Picture Chart (see page 12).

2. Talk about the parts of a plant in "chunks," providing key vocabulary (*leaf, stem, root, seed plant*) and content (*leaves gather sunlight, stem transports food [nutrients] and water from roots to leaves, roots take up nutrients and water from the soil*). As you speak, trace over the corresponding parts with a permanent marker. Do not trace over the words yet.

3. After tracing over the entire picture and providing the vocabulary and content, ask the children to tell what they learned about the parts of a plant. Write the responses on appropriate places on the diagram. Introduce the *Parts of a Plant* Word Cards (pages 41–42) and add them to a pocket chart.

4. Have the children draw a picture of their plant in their *Plants* Science Journal (Observation 9, page 32) and label the parts using the words provided. Ask them to write what each part does. (Assessment)

   ***Parts of a Plant* Home Link**: Have the children complete the *Parts of a Plant* worksheet (page 44) for homework.

Plants (cont.)

Lesson 13: Plant Life Cycle

Review the stages that took place as the seed sprouted and became a plant. Students should be able to tell you these are the stages:

Plant Life Cycle

1. seed   2. seed plant   3. plant

Write the steps of the plant life cycle on the board or on a sheet of chart paper. Explain to the children that the next step happens when the plant makes a flower and then a new seed. Show the children an example of a flower with visible seeds (e.g., sunflower).

Lesson 14: Plants Minibook and Big Book

1. Distribute a *Plants* Minibook (pages 37–39) to each child.

2. Read the book as a class. Discuss the illustrations. Point out key vocabulary (*seed plant, leaf, root, shoot, stem*). Review the *Plants* Word Cards (pages 41–42).

3. Explain what the children are to do at each center.

   Center 1: Read the *Plants* Minibook in small groups with the teacher. Then the children can take turns reading their *Plants* Minibooks to each other.

   Center 2: Color the pictures in the *Plants* Minibook.

   Center 3: Use a yellow crayon to highlight key vocabulary: *alive, air, water,* and *sunlight.* Using a red crayon, circle the high frequency words *are* and *to.* Draw a box around the word *plant* each time you see it. (Assessment)

   Center 4: Listen to a taped version of a related book, such as *The Carrot Seed,* by Ruth Krauss or reread the *Plants* Minibook.

4. At the end of the unit, have each child take home the *Plants* Minibook and "read" it to his or her family.

5. Create an additional class book by enlarging the *Plants* Minibook. Students can take turns coloring the pages. Place the completed *Plants* Big Book in the classroom library.

Plants (cont.)

Culminating Activity: Lima Bean Seeds Experiment

Materials

- 4 small, plastic resealable bags
- lima beans
- permanent marker
- tape
- water

Directions

1. Review with the students the things that plants need to grow (*air, water, sunlight*). Tell them that they are going to do an experiment to see if a seed will grow without air, water, or sunlight.

2. Make four Lima Bean Bags. Using a permanent marker, write on one bag *no air*, on one *no water*, on one *no sunlight,* and on one *air, water, and sunlight.*

3. Add water to the *no air* bag and seal, allowing as little air to remain as possible. Tape the bag to a window with indirect sunlight.

4. Leave the *no water* bag as is and tape to a window with indirect sunlight.

5. Add water to the *no sunlight* bag and seal, and then tape it to the inside of a dark cupboard.

6. Add water to the *air, water, and sunlight* bag, leave it unsealed, and tape it to a window with indirect sunlight.

7. As a group, have the children predict what they think will happen to each seed.

8. Observe the seeds each day to see what changes occur over the next week.

9. At the end of one week, have the students record their observations in their *Plants* Science Journal (Observation 10, page 33). Have the children draw pictures and write sentences, describing the differences between each seed bag and why each did or did not sprout.

Inquiry Skills—Observing, Predicting, Comparing and Contrasting, Designing an Investigation

Extension: Planting Flower Seeds

As an extension, children may plant a few seeds in a Styrofoam cup filled with potting soil. Have the children take the cup home to observe and care for the plants as they grow. These can make a nice gift for parents, grandparents, or someone special.

Assessment Pages: Demonstrating Content Knowledge & Vocabulary

- **Plant Poster** (page 24)—Create a poster to show what a plant needs to grow (*air, sunlight, soil, water*).
- **Parts of a Seed** (page 43)—Cut and paste each *Parts of a Seed* word (*cotyledon, embryo, seed coat*) to the corresponding part of a seed.
- **Parts of a Plant** (page 44)—Cut and paste each *Parts of a Plant* word (*stem, root, leaf, leaf stalk*) to the corresponding part of a plant.

Plant Poster

Culminating Assessment: Plant Poster

Objective: The students will show an understanding of what plants need to grow by creating a poster and labeling plant needs.

Materials (for each student)

- 12" x 18" sheet of white or light blue construction paper (background)
- 4.5" x 6" sheet of yellow or orange construction paper (sun)
- 4.5" x 12" sheet of brown construction paper (soil)
- 4.5" x 6" sheet of medium or dark blue construction paper (water)
- 4.5" x 9" sheet of green construction paper (stem and leaves)
- 4.5" x 6" sheet of light brown construction paper (roots)
- Plant Poster Labels (below)
- glue, scissors, crayons, pencils

Directions for Teacher

1. Photocopy the labels (below) for each child.

2. Distribute materials to each child.

3. Discuss the labels. Explain that the children will need to choose only those labels that show what a plant needs. (Explain that some labels will not be used.)

4. Have the students create a Plant Poster, showing the parts of a plant and what it needs.

5. Have the students cut out the correct labels (*air, soil, sunlight, water*) and glue each to the corresponding part of the Plant Poster.

6. Glue those labels on the proper spots on their posters.

air | sunlight | soil

water | music | night

Science Journal

PLANTS

By: _____

Science Journal

Observation 1: Dissecting Lima Bean Seeds

1. Look at the inside of your lima bean.
   Use a hand lens to look at the different parts.

2. Draw a picture of your lima bean seed.

3. Label each part of the seed.

My Lima Bean

Today is _____ .

Word Bank

| cotyledon | embryo | seed coat |

Science Journal

Observation 2: Lima Bean Bags

1. Draw a picture of the lima bean seeds in your bag.

2. Label each item in your bag.   Use the Word Bank.

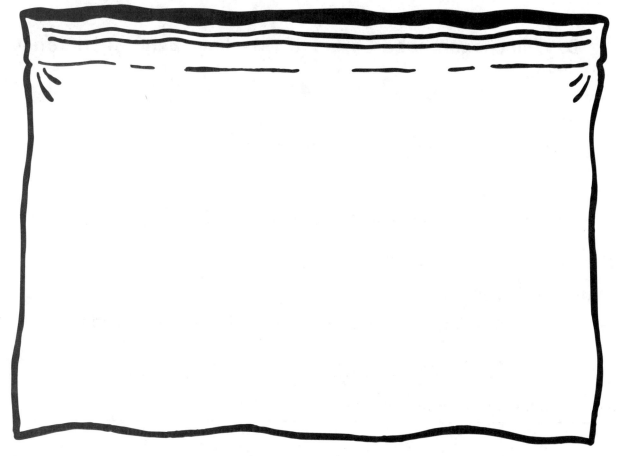

Prediction: *What do you think your lima bean seed will look like tomorrow?*

_____

_____

_____

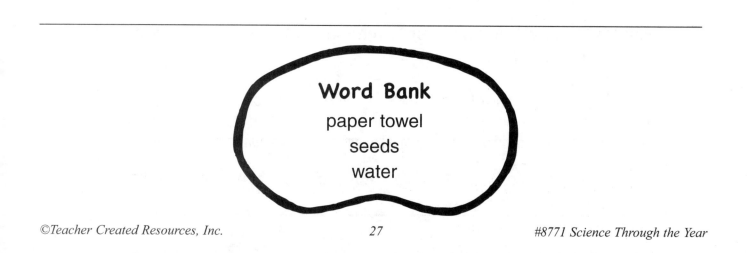

Word Bank

paper towel

seeds

water

Science Journal

Observation 3: Measuring Seeds

1. Draw a picture of a dry lima bean seed.

2. Draw a picture of one of your lima bean seeds today.

Dry lima bean seed	My lima bean seed today

Compare and Contrast

Write two sentences to describe the difference between the two seeds.

_____

_____

_____

_____

_____

Measurement: Use the metric ruler to measure your lima bean seeds.
Record your data below.

Dry seed is _____ mm long   Wet seed is _____ mm long

10   20   30   40   50   60   70   80   90   100 110 120 130 140 150

Observation 4:   Observing Seeds

1. Draw a picture of your lima bean seeds now.

2. Label each part of the seeds you see.

Observation: What is happening to your lima bean seeds?

_____

_____

_____

 # Science Journal

Observation 5: Comparing Seeds

1. Draw a picture of each seed.

2. Label each part of the seeds.

Dry Seed	Seed in Bag (wet)

Compare and Contrast

What is the difference between your dry seed and your wet seed?

_____

_____

What is the same about your dry seed and your wet seed?

_____

_____

Prediction: *What do you think will happen to your seed this week?*

_____

_____

Science Journal

Observations 6-8: Observing Seed Plants

1. Choose one seed plant to observe.
2. Measure the seed plant and record the results.
3. Draw pictures of your seed plant after each observation.

Observation 6 Plant measures _____ mm	
Observation 7 Plant measures _____ mm	
Observation 8 Plant measures _____ mm	

Observation: What is happening to your lima bean plant?

_____

_____

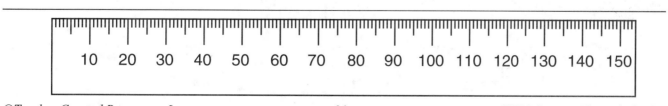

Science Journal

Observation 9: Lima Bean Plant

1. Draw a picture of your lima bean plant.

2. Label each part of the plant.

My Lima Bean

Today is _____ .

Describe: *What job does each plant part do?*

Leaf _____

Roots _____

Stem _____

Word Bank

leaf   roots   stem

Science Journal

Observation 10: Lima Bean Seeds Experiment

Draw	Write
No Air	What happened? _____ _____ _____
No Sunlight	What happened? _____ _____ _____
No Water	What happened? _____ _____ _____
Air, Sunlight, and Water	What happened? _____ _____ _____

Seeds

By: _____

A seed has a baby plant inside.

1

A seed has special parts.

2

The *seed coat* protects the seed.

3

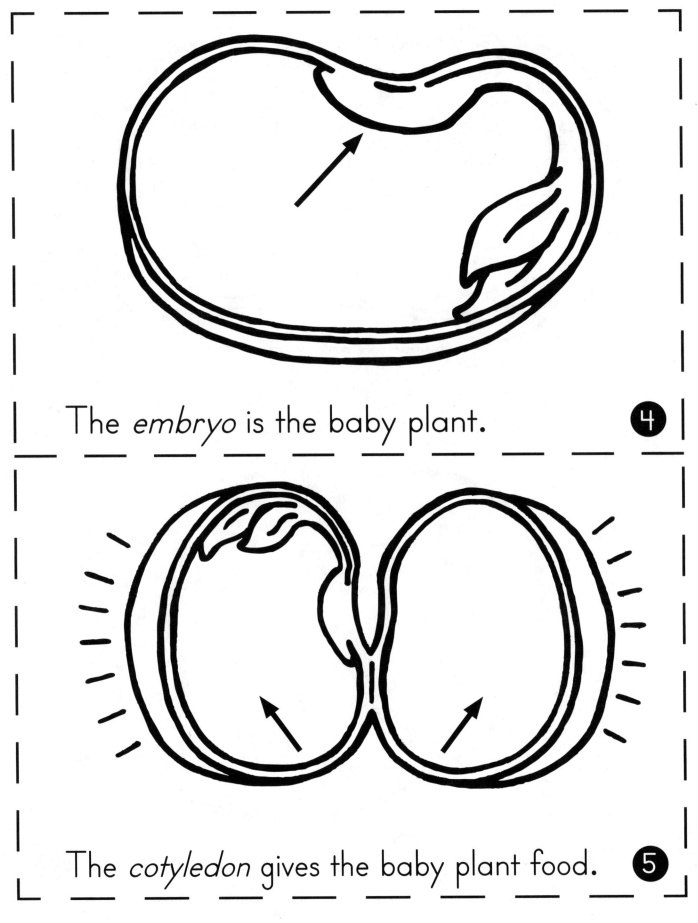

The *embryo* is the baby plant.

4

The *cotyledon* gives the baby plant food.

5

36

Plants

By: _____

Plants are alive.

①

Plants need *air* to grow. **2**

Plants need *water* to grow. **3**

Plants need *sunlight* to grow.

❹

Plants are amazing!

❺

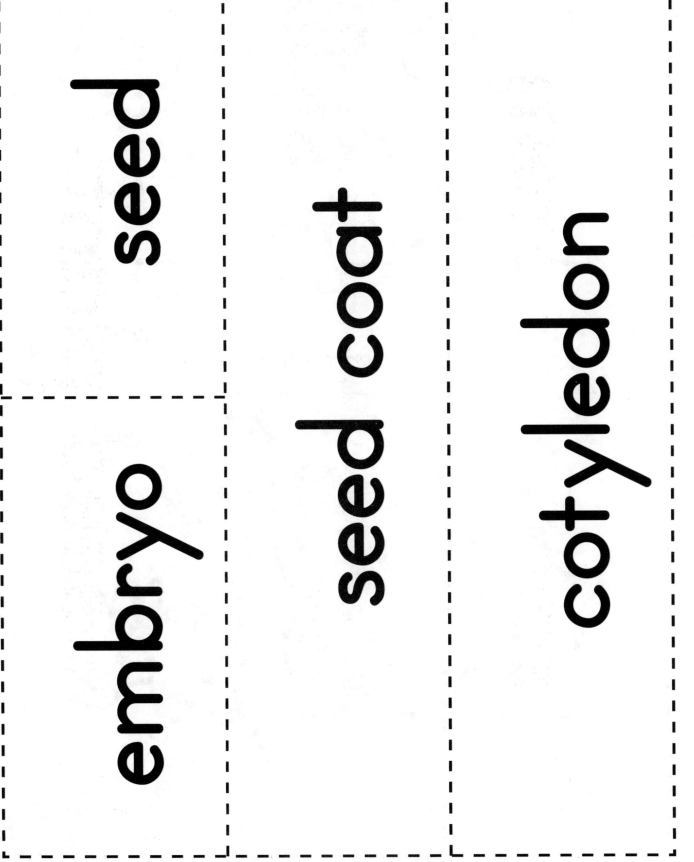

seed

embryo

seed coat

cotyledon

stem

leaf stalk

seed plant

soil

air

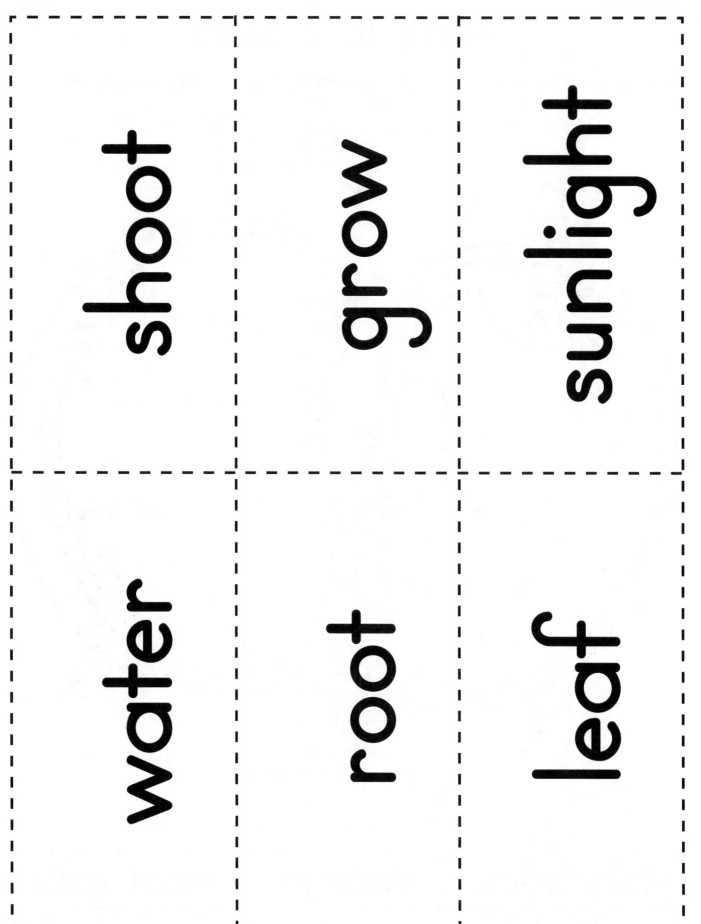

shoot

grow

sunlight

water

root

leaf

Parts of a Seed

Directions: Cut out the word cards below. Glue each card in the correct box.

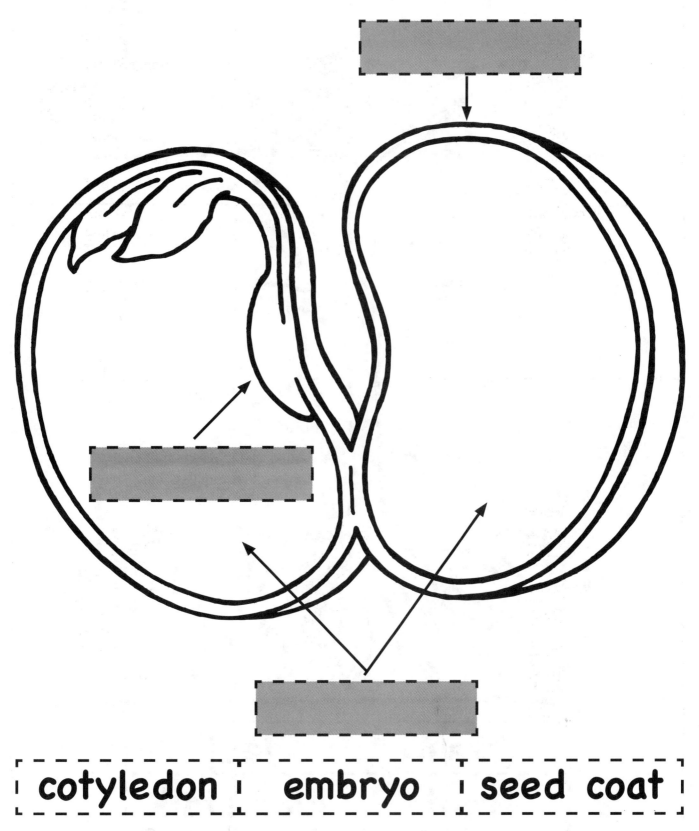

cotyledon   embryo   seed coat

Parts of a Plant

Directions: Cut out the word cards below. Glue each card by the correct plant part.

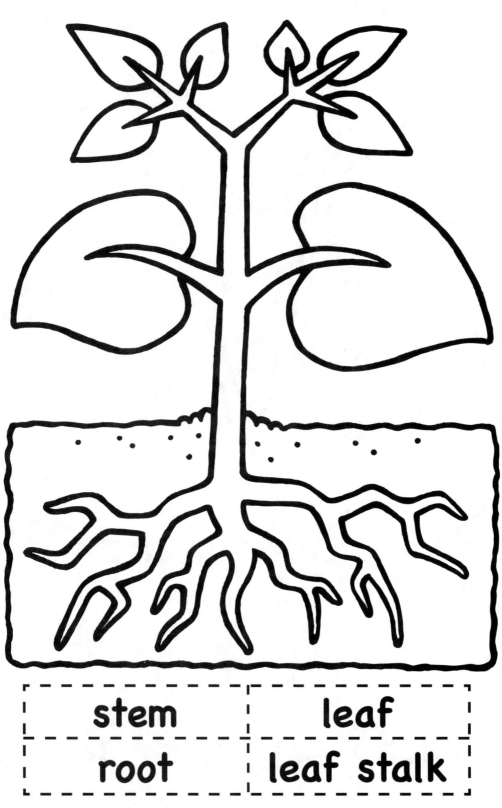

stem	leaf
root	leaf stalk

Parts of an Insect

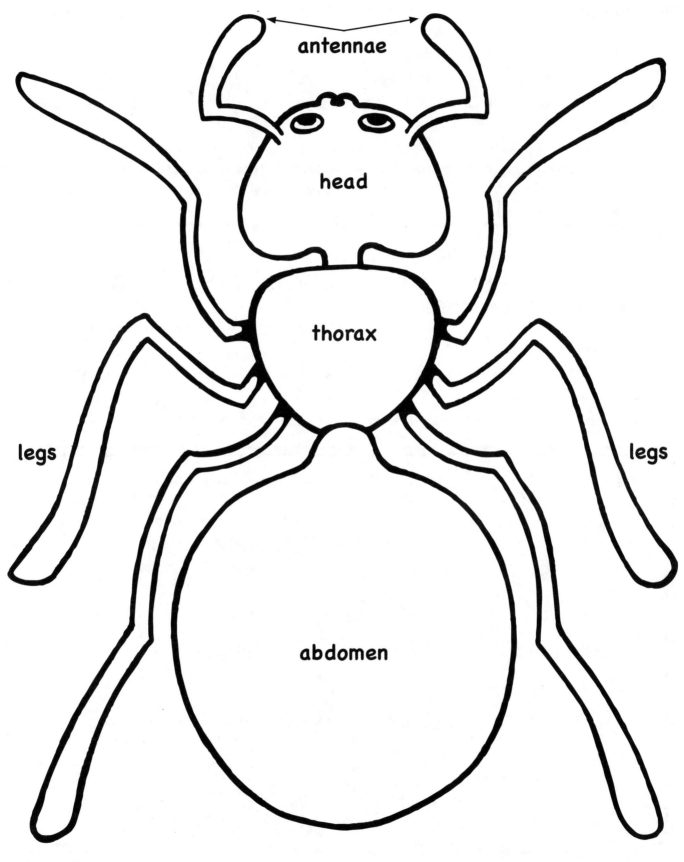

Parts of an Insect Vocabulary

Antennae—insects have two antennae
- detect movements and air vibrations
- vary in size by type of insect
- help the insect touch and smell its environment

Head—first section of an insect that includes insect's eyes, antennae, jaws, and brain
- has two compound eyes, each with hundreds of lenses
  (eyes are called "compound" because they are made up of hundreds of tiny lenses that allow the insect to see in every direction at once)
- uses jaws for biting and chewing food

Thorax—middle section of an insect
- legs are attached to the thorax
- wings are attached to the thorax (some insects have wings [e.g., ladybird beetles], some do not [e.g., harvester ants])

Abdomen—includes "gut," which is like a stomach; air sacs for breathing; and heart
- some insects have a stinger at the end of their abdomen (e.g., bees)

Legs—insects have legs
- has six jointed legs attached to the thorax

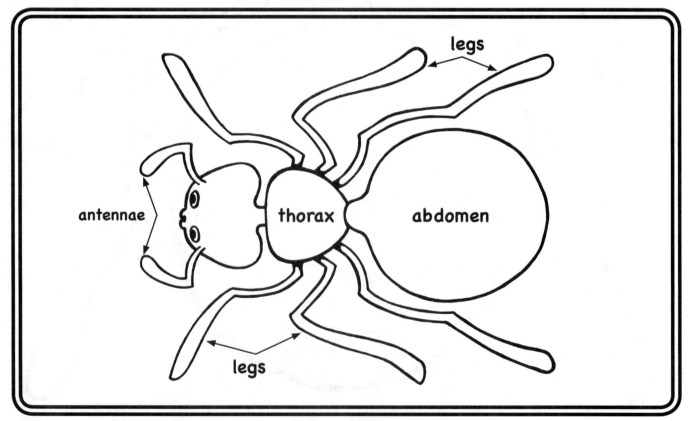

Insect Life Cycle

Grain Beetle

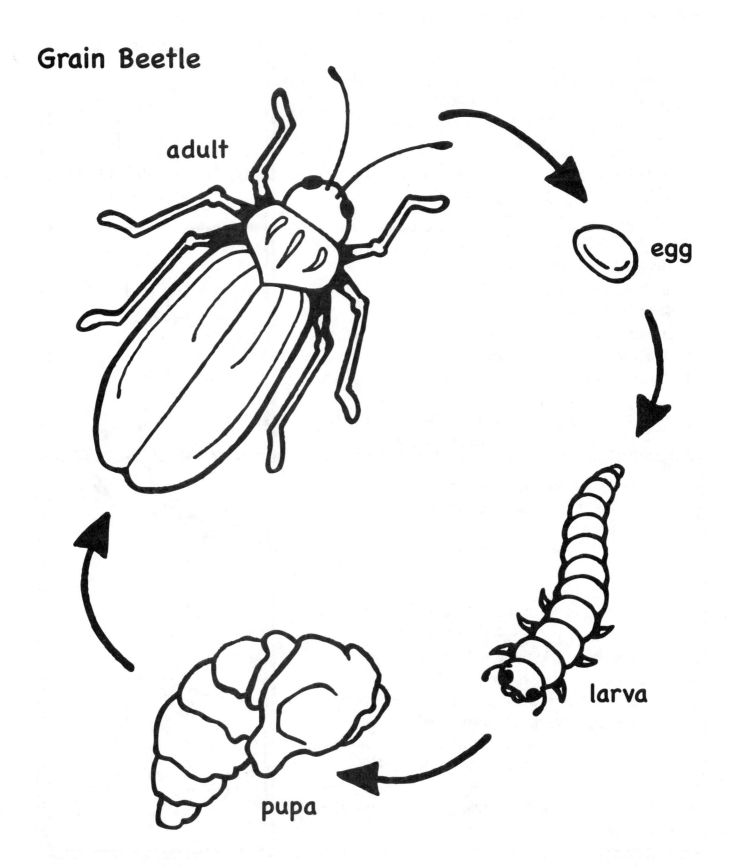

Insect Life Cycle Vocabulary

Complete Metamorphosis—metamorphosis (change) with four stages

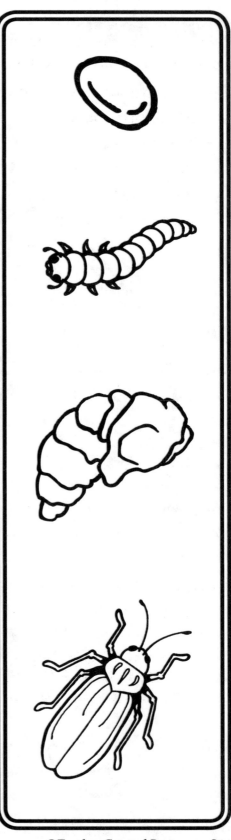

Egg—first stage of complete metamorphosis
- laid by a female adult insect
- very small (1 mm), white, oval-shaped
- hatches in 1–2 weeks

Larva—second stage of complete metamorphosis
- wormlike creature with three pairs of legs; two antennae; and head, thorax, and abdomen
- eats rotten grain and food (is a decomposer)
- cream-colored, harmless
- is a larva for 2–4 weeks

Pupa—third stage of complete metamorphosis
- doesn't eat or crawl, but twitches sometimes
- encased in light brown sac
- looks like it is dead, but it is not
- is a pupa for 1–3 weeks
- inside, the whole insect is being "rebuilt"

Adult—fourth stage of complete metamorphosis
- is white when it emerges from the pupa, then turns dark brown to black
- eats and lives in grain
- has all the parts of an insect: head, thorax, abdomen, six legs, and wings
- adult grain beetles (also known as darkling beetles or stink bugs) mate, then the female lays up to 500 eggs
- dies soon after the eggs are laid and the cycle begins all over again

- -

Date _____

Dear Parent,

We will be starting a unit on insects soon. There are a few items we need for the unit. Please provide the item circled below. If you are unable to provide the item, please let me know as soon as possible.

- container of regular oatmeal (not quick cooking)
- 30 plastic spoons
- 30 paper plates
- 30 plastic cups with lids
- apples (any type)
- large box of raisins

Please send your item to school with your child by _____ .

Thank you for your help with this unit!

Sincerely,

- -

Date _____

Dear Parent,

We have just started a science unit on insects! We will be watching insects as they go through the stages of metamorphosis. Please ask your child what kinds of insects he or she is learning about.

At the end of the unit, students will have the option to take home their insects, as well as the *Insects* and *Insect Life Cycle* Minibooks they have been reading in class. Ladybird beetles can be set free outdoors after a day or two; mealworms live about a week and then should be composted. Please ask your child to "read" his or her books to you and tell you about the changes that insects go through during their life cycle.

Sincerely,

- -

Insects

 Background Information

Animals have predictable life cycles. Grain beetles (mealworms) and ladybird beetles go through complete metamorphosis. The following are stages of complete metamorphosis: egg, larva, pupa, and adult. Insects have six legs and three body parts: head, thorax, and abdomen. Studying insects provides students an opportunity to observe changes in a life cycle over time. Students will discover that animals need air, water, and food to survive. Students use inquiry skills, such as observation, measurement, prediction, comparing/contrasting, and recording data, when studying insects.

It should take about one month for the mealworms to go through their entire life cycle. It should take about two weeks for the ladybird beetle larvae to pupate and emerge as adult ladybird beetles. Harvester ants can live for months in their ant farm if properly cared for.

 Unit Preparation

1. Purchase the following items or ask for donations (see letter on page 49):

   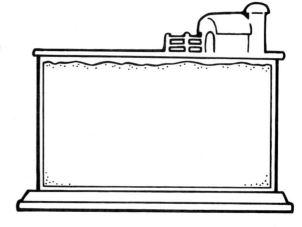

   - ant farm (order harvester ants*)
   - ladybird beetle habitat (order larvae**)
   - container of regular oatmeal (not quick cooking)
   - 30 plastic spoons
   - 30 paper plates
   - 30 plastic cups with lids
   - spray bottles
   - small apples (any type)
   - large box of raisins
   - plastic, disposable rectangular container with lid
   - small hand lenses
   - 100 small mealworms (from pet store or bait shop)
   - black and white construction paper

* Ant farms can be ordered through online companies such as **Carolina Biological Supply** (*www2.carolina.com*). You may purchase an **Uncle Milton's**® ant farm at a discount department or toy store. When purchasing an ant farm, you must order the ants separately by sending in the certificate that comes with the farm.

You may wish to substitute crickets for the harvester ants. Crickets can be purchased at a pet store. House them in a small plastic container with a ventilated lid (available at pet store). Feed them cricket food (available at pet stores). Add a piece of wet, paper egg carton to the cricket container so they can crawl on it and draw moisture from it. After the children study them, crickets can be fed to pets, such as reptiles, or let go.

** **Insect Lore** ladybird beetle habitats are sold at discount department and toy stores. You must send in the certificate that comes with the habitat to receive the ladybird beetle larvae. Or buy adult ladybird beetles at a garden store—omit the lesson plans on ladybird beetle larvae and do the lessons on adult insects instead—house them in their own small plastic container with ventilated lid (available at pet store); feed soaked raisins; after the children study them, ladybird beetles can be released.

Insects *(cont.)*

 ## Unit Preparation *(cont.)*

2. Copy and cut out the *Parts of an Insect* Word Cards (page 84) and *Insect Life Cycle* (page 85).

3. Copy and assemble the *Insects and Insect Life Cycle* Minibooks (pages 78–83).

4. Copy and assemble the *Insects* Science Journals (pages 65–77).

5. Reproduce the *Insects* Assessment sheets (pages 63–64).

6. Use chart paper or butcher paper to create the Inquiry Chart for Lesson 1 (see page 52), as well as the whole-class charts used throughout the unit.

7. Prepare the mealworms by adding the oatmeal to the container, half an apple, and the mealworm larvae. Keep the lid off for good air circulation until the adult beetles form.

   Teacher Tip: The mealworms will live in and eat the oatmeal. They will also eat the apple half; it serves as a source of moisture. Change the apple every few days to prevent mold. Mealworms require nothing else and are very easy to care for.

Literature Links

The Very Quiet Cricket by Eric Carle

The Very Grouchy Ladybug by Eric Carle

The Very Lonely Firefly by Eric Carle

The Very Hungry Caterpillar by Eric Carle

Mealworms: Raise Them, Watch Them, See Them Change by Adrienne Mason

A Mealworm's Life by John Himmelman

Mealworms (Life Cycles) by Donna Schaffer

The Magic School Bus Gets Ants in the Pants by Joanna Cole

The Magic School Bus Explores the World of Bugs by Nancy White

The Magic School Bus Inside a Beehive by Joanna Cole

Insects *(cont.)*

Lesson 1: Introducing the Insect Unit

Ask the children what kinds of insects they have seen at home in their backyard or walking in the neighborhood. Ask them to turn to a neighbor and tell him or her about the insects they have seen. Allow a few students to share their examples with the whole class. Tell them that today we are starting a science unit about insects, but first, we need to find out what we already know. Show the class the Insects Inquiry Chart (see below) and ask them to help you complete each section of the chart.

Note: *Color-code each column of the chart using bright markers such as red, blue, and green.*

Insects Inquiry Chart

What Do You Know About Insects?	Insects We've Seen	What Do You Want to Learn About Insects?	How Can We Find Out?

The inquiry chart serves as an assessment. It demonstrates what the children already know, what their misconceptions may be, and what they are wondering about. It is an excellent springboard for future lessons. Inquiry charts can provide teachers with needed information when planning lessons.

The Sample Insects Inquiry Chart (below) gives an idea of the kinds of responses first grade children might give.

Sample Insects Inquiry Chart

What Do You Know About Insects?	Insects We've Seen	What Do You Want to Learn About Insects?	How Can We Find Out?
Insects are bugs. Bees sting you. Ladybugs fly. If you touch a ladybug, it won't poison you.	Spiders Ants Ladybugs Bees Wasps	Is a butterfly an insect? Is an ant an insect? How do you know if it's an insect? How many eggs do they lay at once?	Read books. Go to the library. Check the Internet. Ask our parents. Ask the teacher. Look at insects!

Note: *At the end of each unit, you can have the children help you complete a chart of "What We Learned" as a whole group culminating assessment.*

Insects *(cont.)*

Lesson 2: Harvester Ants

Introduce the class to the harvester ants. Explain that the ants must stay in their ant farm and that the students will need to take care of them. Ask the children if they have ever seen ants at home or on a walk. There should be many responses. Ask the children what they think the ants will need to survive. Encourage children to think of examples from home, perhaps a time when someone left out food on the kitchen counter and ants came or when ants turned up in the bathroom to find water. Record the students' responses on chart paper or on the board. After the class has brainstormed a list of ideas, explain to them that the ants need air, water, and food to survive. Explain that the ants have plenty of air in the ant farm, but the children will need to give the ants a bit of food and a few drops of water every other day. Have the children draw a picture of the ants in the ant farm in their *Insects* Science Journals and ask them to predict what will happen to the ant farm over the next week (Observation 1, Prediction 1, page 66).

Inquiry Skills—Observation, Prediction

Lesson 3: *Parts of an Insect* Picture Chart

1. Gather the children together on the rug near the *Parts of an Insect* Picture Chart (see page 45) to teach content and vocabulary. Show the children the chart paper with the pre-drawn pencil diagram of the parts of an insect. (Be sure to write your notes lightly in pencil next to each part of the diagram.)

2. Talk about the parts of an insect in "chunks," providing key vocabulary (*antennae, head, thorax, abdomen, six legs*) and content (*insect uses antennae to smell; head has the eyes and brain; abdomen has the gut, heart, and stomach*). As you speak, trace over the corresponding parts with a permanent marker. Do not label the diagram yet.

   Teacher Note: Describe what you are doing as you draw. Mention the curving lines, larger and smaller shapes, etc. Your descriptions should help students feel more comfortable with their own Science Journal illustrations.

3. After tracing over the entire picture and providing the vocabulary and content, ask the children to tell what they learned about the parts of an insect.

4. Write the children's responses on appropriate places on the chart (i.e., labeling parts).

5. Introduce the *Parts of an Insect* Word Cards (page 84) and add them to a pocket chart.

6. Display the completed *Parts of an Insect* Picture Chart in the classroom for a reference tool.

 Parts of an Insect **Home Link:** Have the children complete the *Parts of an Insect* worksheet (page 86) for homework.

Insects (cont.)

Lesson 4: Ladybird Beetle Larvae

Introduce the class to the ladybird beetle larvae. Explain that these are ladybird beetle larvae, or "baby" ladybugs. Explain that most people call these insects ladybugs, but their real name is *ladybird beetle*. Ask the class, *How are these different from or the same as the adult ladybird beetles you have seen?* Record student responses on chart paper using a Venn diagram comparing ladybird beetle larvae and adult ladybird beetles. Explain to the students that they will observe ladybird beetles each day to see if there are any changes. Have the children draw a picture of the ladybird beetle larvae in their *Insects* Science Journals (Observation 2, page 67).

Inquiry Skill—Compare and Contrast

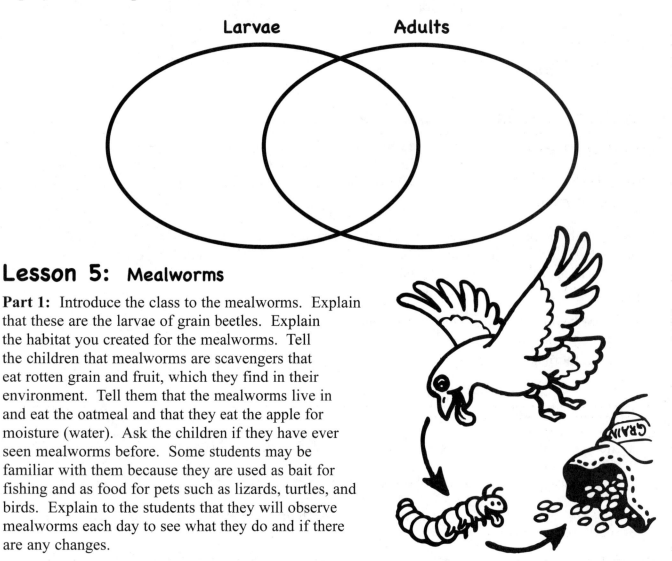

Larvae   **Adults**

Lesson 5: Mealworms

Part 1: Introduce the class to the mealworms. Explain that these are the larvae of grain beetles. Explain the habitat you created for the mealworms. Tell the children that mealworms are scavengers that eat rotten grain and fruit, which they find in their environment. Tell them that the mealworms live in and eat the oatmeal and that they eat the apple for moisture (water). Ask the children if they have ever seen mealworms before. Some students may be familiar with them because they are used as bait for fishing and as food for pets such as lizards, turtles, and birds. Explain to the students that they will observe mealworms each day to see what they do and if there are any changes.

Part 2: Explain to the children that they will work in pairs to observe mealworms. Provide each pair with a paper plate, a hand lens, and 2–3 mealworms. After they have observed their mealworms, they will draw a picture of their mealworms and write about them in their *Insects* Science Journals (Observation 3, Prediction 2, page 68).

Inquiry Skills—Observation, Predicting

Insects *(cont.)*

Lesson 5: Mealworms *(cont.)*

The following activities are fun to have the children try with the mealworms after they have completed Part 2 (see page 54).

Mealworm Races

Materials (for each pair)

- 2 mealworms
- paper plate

Directions

1. Place two mealworms on one side of the paper plate.
2. Draw a line on the opposite side of the plate for the "finish line."
3. Observe the mealworms to see which one "races" across the finish line first!
4. Record the results in your *Insects* Science Journals (page 69).

Light or Dark?

Materials (for each pair)

- 2 mealworms
- paper plate
- 4 1/2″ x 6″ (11.4 cm x 15.2 cm) piece of white construction paper
- 4 1/2″ x 6″ (11.4 cm x 15.2 cm) piece of black construction paper

Directions

1. Place each paper on one side of the paper plate.
2. Place two mealworms on the paper plate.
3. Observe the mealworms to see if they crawl to the light or dark side.
4. Record the results in your *Insects* Science Journals (page 70).

Wet or Dry?

Materials (for each pair)

- 2 mealworms
- paper plate
- access to spray bottle filled with water

Directions

1. Using a spray bottle, wet one side of the paper plate.
2. Place two mealworms on the paper plate.
3. Observe the mealworms to see if they crawl to the wet or dry side.
4. Record the results in your *Insects* Science Journals (page 71).

Insects (cont.)

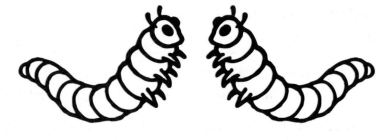

Lesson 6: Measuring Mealworms

Materials (for each pair)

- paper plate
- plastic spoon
- 2 mealworms

Directions

1. Use the metric ruler at the bottom of the *Insects* Science Journal (page 72) to measure each mealworm.

2. Record each length in the *Insects* Science Journal (Observation 4, page 72).

Inquiry Skill—Measurement

Lesson 7: Examining Insect Larvae

Before the ladybird beetle and grain beetle larvae have pupated, have the students work in small groups to examine the insects closely.

Materials (for each group of four students)

- 2 small plastic cups, one with 2–3 mealworms, one with 2–3 ladybird beetle larvae
- hand lens
- paper plate
- plastic spoon

Directions

1. Have each Materials Manager collect the materials and bring them back to his or her group.

2. Instruct the children to carefully tip their cup and dump the mealworms onto the paper plate. They can use the hand lens to examine the larvae and the spoon to move them around. It's okay for the children to touch the mealworms with their hands, as long as they wash them after the activity. Ask the students to think about the *Parts of an Insect* Picture Chart (see page 45) and look for the various parts in the larvae. Have the students use the spoon to carefully put the mealworms back into the cup.

3. Students repeat Step 2 using ladybird beetle larvae.

Insects *(cont.)*

Lesson 7: Examining Insect Larvae *(cont.)*

Directions *(cont.)*

4. Ask the Reporter from each group to share the group's observations.

5. Students work independently to draw a picture of each type of insect larvae (ladybird beetle, mealworm) in their *Insects* Science Journal and label the parts they see (Observation 5, page 73).

   Note: On larvae, the thorax and abdomen are there but difficult to see. It's helpful to show students a diagram from a children's science book such as Mealworms *by Adrienne Mason.*

6. Complete the Compare/Contrast Insect Larvae Chart (below) as a class.

Inquiry Skills—Observation, Compare and Contrast

Compare/Contrast Insect Larvae Chart

Type of Larva	Number of Legs?	(visible) Head?	(visible) Thorax?	(visible) Abdomen?
Grain beetle	6	yes	yes	yes
Ladybird beetle	6	yes	yes	yes

Lesson 8: Ongoing Observation

Students should have the opportunity to observe the insect larvae each day. The teacher can check the mealworms each day to check for changes. On days when changes occur, have the children carefully look at the mealworms and record their observations in their *Insects* Science Journals (Observations 6–8, page 74). Remind the students to look for insect body parts. This can easily be done during rug time or during calendar time when the teacher can keep a tally of how many days they have had the insect larvae.

Inquiry Skill—Observation

Science Journal

Observations 6–8: Observing the Mealworms

1. Draw pictures of the mealworms.

2. Write a sentence about each picture. Tell what your mealworms are doing.

Observation 6

_____
_____
_____
_____

Observation 7

_____
_____
_____
_____

Observation 8

_____
_____
_____
_____

Insects *(cont.)*

Lesson 9: Literature Connection

Introduce the story *The Very Quiet Cricket* by Eric Carle to the class. This book is an excellent one to use because the illustrations clearly show the parts of several types of insects, including a cricket, locust, praying mantis, spittlebug, cicada, bumblebee, and dragonfly. The story is also interesting to children and the language is predictable. Do a "picture walk" first, and ask the children to predict what will happen at various points in the story. Then read the story to the class. Have the children help you complete a Story Elements Chart and a Plot Analysis Chart (see below). (Whole Class Assessment)

Sample Story Elements Chart

Setting	Characters
Morning, grass	Locust
Day, grass	Praying Mantis
Day, under apple tree	Worm
Day, tree branch	Cicada
Afternoon, flowers	Bumblebee
Night, field	Mosquitoes
Night, air	Luna Moth

Sample Plot Analysis Chart

Beginning	Middle	End
A little cricket emerged from a tiny egg.	He met all kinds of other insects, but he couldn't chirp to answer them.	That night he met a little female cricket, and he chirped for the first time!

Insects *(cont.)*

Lesson 10: *Insects* Minibook and Big Book

1. Distribute an *Insects* Minibook (pages 78–80) to each child.

2. Read the book as a class. Discuss the illustrations. Point out key vocabulary (*head, thorax, abdomen, legs*). Review the *Parts of an Insect* Word Cards (page 84).

3. Explain what the children are to do at each center.

   Center 1: Read the *Insects* Minibook in small groups with the teacher. Then independently read the *Insects* Minibook several times to develop fluency.

   Center 2: Color the pictures in the *Insects* Minibook.

   Center 3: Use a yellow crayon to highlight key vocabulary: *head, thorax, abdomen, legs*. Using a red crayon, circle the high frequency words *an* and *has*. Draw a box around the word *insect* each time you see it. (Assessment)

   Center 4: Listen to a taped version of *The Very Quiet Cricket* or the *Insects* Minibook.

4. At the end of the unit, each child will take home the *Insects* Minibook and "read" it to his or her family.

5. Create an additional class book by enlarging the *Insects* Minibook. Students can take turns coloring the pages. Place the completed *Insects* Big Book in the classroom library.

Lesson 11: Examining Adult Insects

After the ladybird beetle and grain beetle larvae have pupated and emerged as adult insects, have the students work in small groups to closely examine the two types of insects.

Materials (for each group of four students)

- 2 small plastic cups, one with 2–3 adult grain beetles, one with 2–3 adult ladybird beetles
- hand lens
- large sheet of chart paper
- markers

Directions

1. Have each Materials Manager collect the materials and bring them back to his or her group.

2. Instruct the children to carefully pick up the cups and use the hand lens to examine the insects. Ask them to look at the *Parts of an Insect* Picture Chart (see page 45) and look for the various parts on the insects.

Insects (cont.)

Lesson 11: Examining Adult Insects (cont.)

Directions (cont.)

3. While the groups are working, the teacher can bring the ant farm from group to group to show them the harvester ants. (The ants cannot be removed from the ant farm.)

4. Ask the Reporter from each group to share the group's observations.

5. Students work independently to draw a picture of each type of insect (ladybird beetle, grain beetle, harvester ant) in their *Insects* Science Journals and label the insect parts (Observations 9–11, pages 75–77).

6. Introduce the Compare/Contrast Adult Insects Chart (see sample below) to the class.

Inquiry Skills—Observation, Compare and Contrast

Sample Compare/Contrast Adult Insects Chart

Type of Insect	Number of Legs?	Wings or No Wings?	Three Body Parts?	Antennae?
Grain beetle	6	Has wings	Yes	Yes
Ladybird beetle	6	Has wings	Yes	Yes
Harvester ant	6	No wings	Yes	Yes

Insects *(cont.)*

Lesson 12: *Insect Life Cycle* Picture Chart

1. Gather the children together on the rug near the *Insect Life Cycle* Picture Chart (see page 47) to teach content and vocabulary. Show the children the chart paper with the pre-drawn pencil diagram of the parts of an insect. (Be sure to write your notes lightly in pencil next to each part of the diagram.)

2. Talk about the four stages of complete metamorphosis in "chunks," providing key vocabulary (*egg, larva, pupa, adult*) and content (*egg is laid by a female adult insect and hatches in 1–2 weeks; larva is a wormlike creature with three pairs of legs, two antennae, head, thorax, and abdomen; the pupa has a case—inside the insect is being "rebuilt"; adult has all parts of an insect and dies soon after eggs are laid*). As you speak, trace over the corresponding parts with a permanent marker. Do not label the diagram yet.

   Teacher Note: Describe what you are doing as you draw. Mention the curving lines, larger and smaller shapes, etc. Your descriptions should help students feel more comfortable with their own Science Journal illustrations.

3. After tracing over the entire picture and providing the vocabulary and content, ask the children to tell what they learned about the life cycle of a grain beetle.

4. Write the children's responses on appropriate places on the chart (i.e., labeling stages).

5. Introduce the *Insect Life Cycle* Word Cards (page 85) and add them to a pocket chart.

6. Display the completed *Insect Life Cycle* Picture Chart in the classroom for a reference tool.

 Insect Life Cycle **Home Link:** Have the children complete the *Insect Life Cycle* Worksheet (page 87) for homework.

Lesson 13: *Insect Life Cycle* Minibook and Big Book

1. Distribute an *Insect Life Cycle* Minibook (pages 81–83) to each child.

2. Read the book as a class. Discuss the illustrations. Point out key vocabulary (*egg, larva, pupa, adult*). Review the *Insect Life Cycle* Word Cards (page 85).

3. Explain what the children are to do at each center.

   Center 1: Read the *Insect Life Cycle* Minibook with a partner, taking turns reading each page.

   Center 2: Color the pictures in the *Insect Life Cycle* Minibook.

   Center 3: Use a yellow crayon to highlight key vocabulary: *egg, larva, pupa, adult.* Using a red crayon, circle the high frequency words *it* and *an.* Draw a box around the word *cycle* each time you see it. (Assessment)

   Center 4: Listen to a taped version of *The Very Hungry Caterpillar* by Eric Carle or the *Insect Life Cycle* Minibook.

4. At the end of the unit, each child will take home the *Insect Life Cycle* Minibook and "read" it to his or her family.

5. Create an additional class book by enlarging the *Insect Life Cycle* Minibook. Students can take turns coloring the pages. Place the completed *Insect Life Cycle* Big Book in the classroom library.

Insects *(cont.)*

Culminating Activity: Taking Home the Mealworms and Ladybird Beetles

Materials (for each student)

- 2 plastic cups with lids
- 2 grain beetles
- 2 ladybird beetles
- grass

- raisin (soaked)
- oatmeal
- small piece of apple

Directions

1. Place the grain beetles, a little bit of oatmeal, and a small piece of apple in a plastic cup.

2. Place the ladybird beetles, a bit of grass, and a soaked raisin in another plastic cup.

3. Observe your ladybird beetles at home for a few days; then release the ladybird beetles.

4. Observe your grain beetles in the cup for a week. Add a fresh piece of apple every few days.
   Inquiry Skill—Observation

Assessment Pages: Demonstrating Content Knowledge and Vocabulary

- **Life Cycle Circle** (pages 63–64)—Show the correct order of the insect life cycle by gluing each stage onto the correct part of a paper plate.

- **Parts of an Insect** (page 86)—Cut and paste each *Parts of an Insect* word (*thorax, head, leg, eye, abdomen, antenna*) to the corresponding part of the insect.

- **Insect Life Cycle** (page 87)—Cut and paste each *Insect Life Cycle* word (*adult, pupa, egg, larva*) to the corresponding part of the chart.

Insects *(cont.)*

Culminating Assessment: Life Cycle Circle

Objective: Show the correct order of the insect life cycle by gluing each stage onto the correct part of a paper plate.

Materials (for each student)

- Large white paper plate (divided into four sections with a marker)
- Life Cycle Circle Pictures (page 64)
- glue, scissors, crayons, pencil

Directions

1. Distribute the materials to each child.

2. Ask the students to color and cut out each stage of the Insect Life Cycle Pictures (page 64).

3. Have the students show the correct order of the insect life cycle by gluing each stage onto the correct part of the plate (see sample below).

Life Cycle Circle Pictures

Directions

1. Color and cut out each picture.

2. Glue the pictures on a paper plate in the correct order.

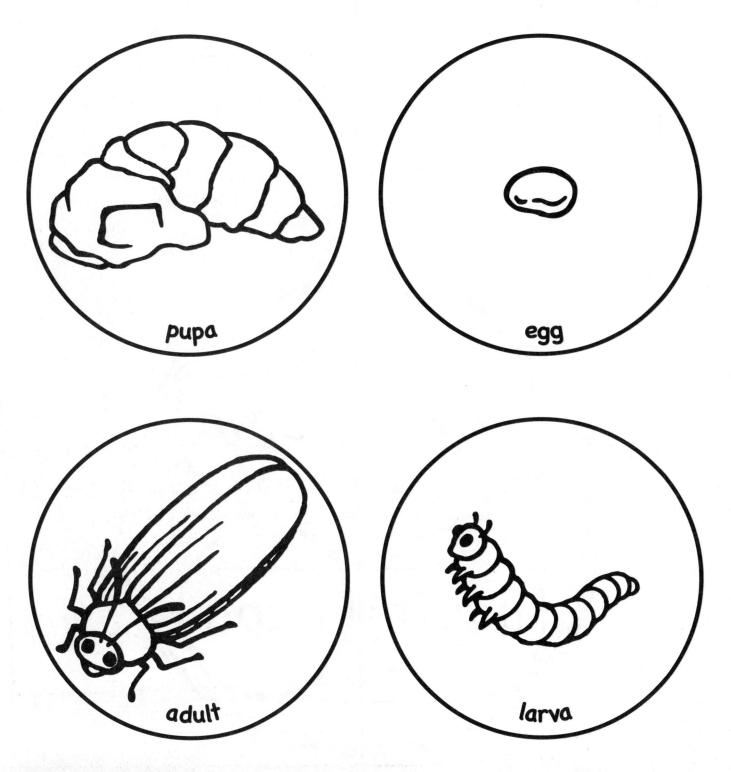

pupa

egg

adult

larva

Science Journal

Insects

By: _____

Science Journal

Observation 1: Harvester Ants

1. Draw a picture of a harvester ant.

2. Label each part of the ant.

Observation: What do you see the ants doing in the ant farm?

_____

_____

Prediction 1: What do you think the ants will do this week?

_____

_____

Word Bank

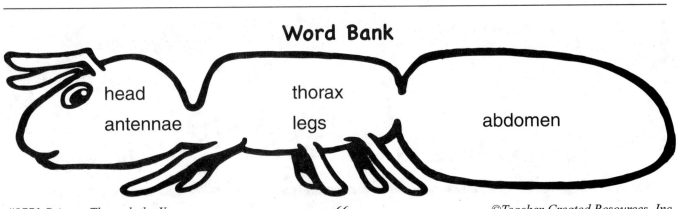

head

antennae

thorax

legs

abdomen

Science Journal

Observation 2: Ladybird Beetle Larvae

1. Draw a picture of a ladybird beetle larva.

2. Draw a picture of what the larva eats.

3. Label each part of the larva.

Observation: What are the ladybird beetles doing in their habitat?

Word Bank

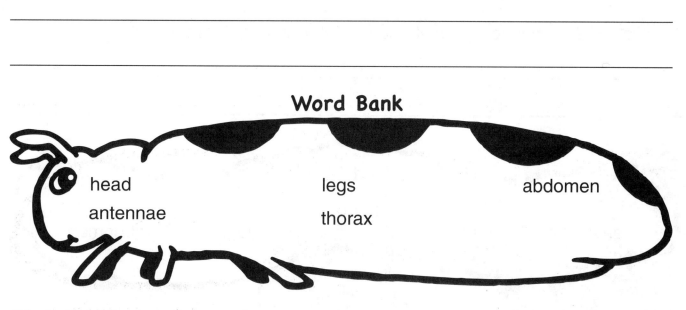

head   legs   abdomen

antennae   thorax

Science Journal

Observation 3: Mealworms

1. Draw a picture of a mealworm.

2. Draw a picture of what the mealworm eats.

3. Label each part of the mealworm.

Prediction 2: What do you think the mealworms will do this week?

_____

_____

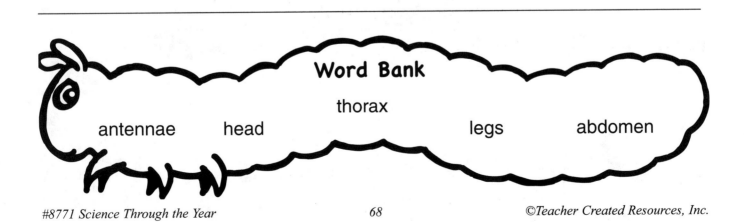

Word Bank

antennae   head   thorax   legs   abdomen

Science Journal

Mealworm Races

1. Draw your mealworms.

2. Write about the Mealworm Race.

Observation: What happened during the Mealworm Race?

_____

_____

_____

Science Journal

Light or Dark?

1. Draw your mealworms.

2. Write about your mealworms.

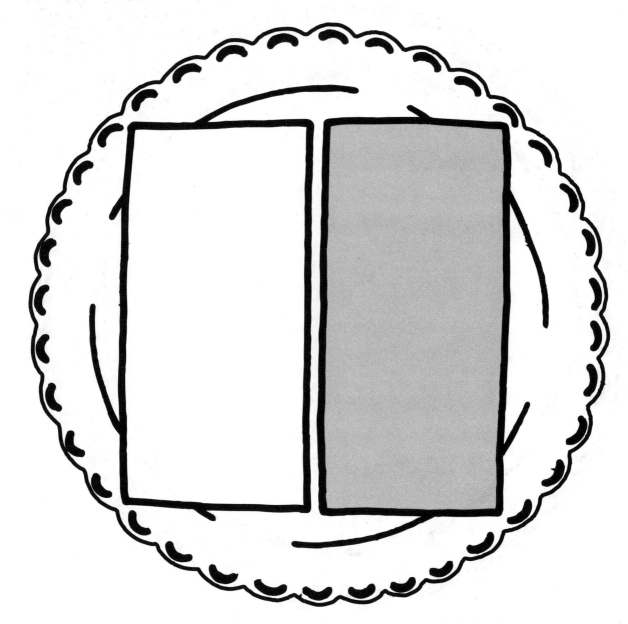

Observation: Do the mealworms like light or dark best?

_____

_____

_____

Science Journal

Wet or Dry?

1. Draw your mealworms.

2. Write about your mealworms.

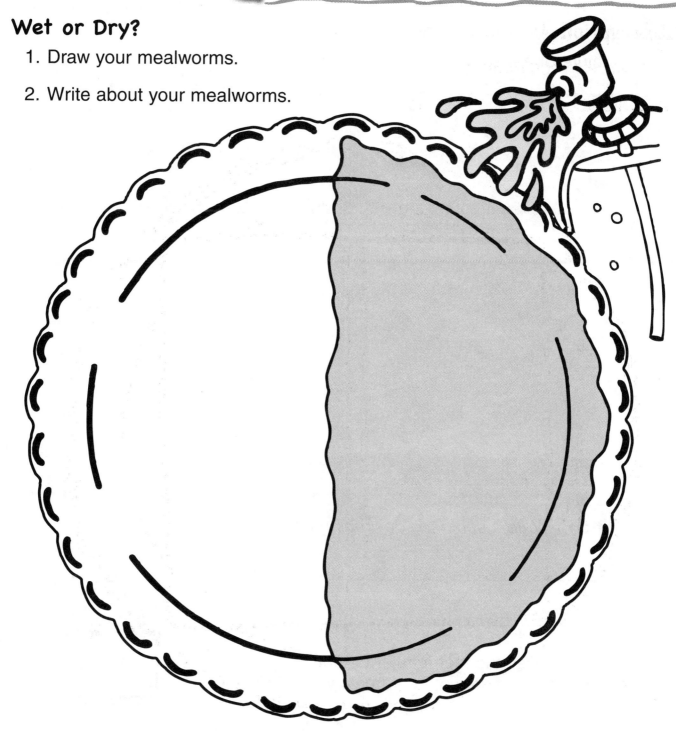

Observation: Do the mealworms like wet or dry best?

_____

_____

_____

Science Journal

Observation 4: Measuring Mealworms

Draw a picture of a mealworm that is moving.

Observation: *Write a sentence describing how the mealworms move.*

_____

_____

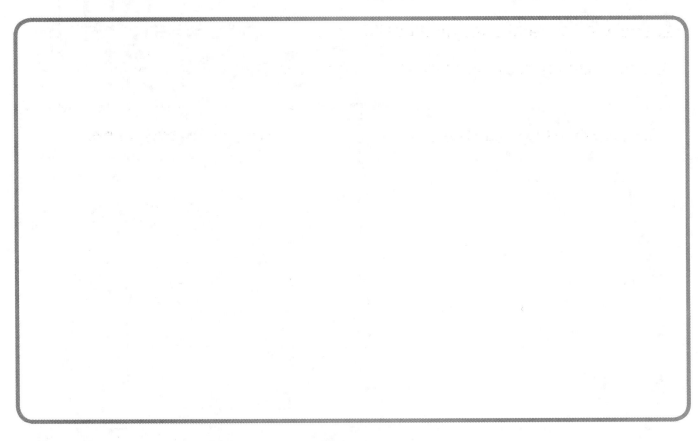

Measurement: Use the metric ruler to measure your mealworms. Record your data below.

Mealworm 1 is _____ mm long   **Mealworm 2** is _____ mm long

Science Journal

Observation 5: Examining Insect Larvae

1. Draw a picture of a mealworm larvae.

2. Draw a picture of a ladybird beetle larvae

3. Label each part of the mealworm and ladybird beetle.

Mealworm (Grain Beetle) Larva	Ladybird Beetle Larva

Compare and Contrast

What is the difference between the grain beetle and ladybird beetle larvae?

_____

_____

What is the same about the grain beetle and ladybird beetle larvae?

_____

_____

Word Bank

head	abdomen	legs	thorax	antennae

Science Journal

Observations 6–8: Observing the Mealworms

1. Draw pictures of the mealworms.

2. Write a sentence about each picture. Tell what your mealworms are doing.

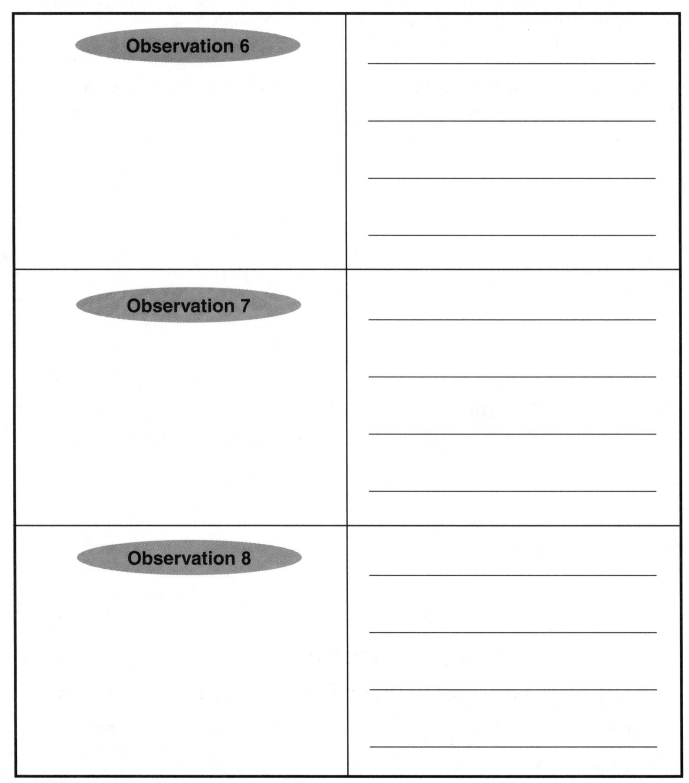

Observation 6

_____

_____

_____

_____

Observation 7

_____

_____

_____

Observation 8

_____

_____

_____

_____

Science Journal

Observation 9: Adult Harvester Ant

1. Draw a picture of an adult harvester ant.

2. Label each part of the ant.

Word Bank

head   antennae   thorax   legs   abdomen

Science Journal

Observation 10: Adult Ladybird Beetle

1. Draw a picture of an adult ladybird beetle.

2. Label each part of the ladybird beetle.

Word Bank

head   antennae   thorax   legs   abdomen

 # Science Journal

Observation 11: Adult Grain Beetle

1. Draw a picture of an adult grain beetle.

2. Label each part of the grain beetle.

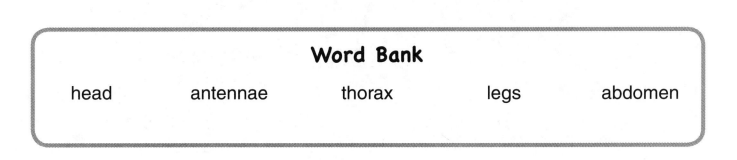

Word Bank

| head | antennae | thorax | legs | abdomen |

Insects

By: _____

An insect has six legs.

1

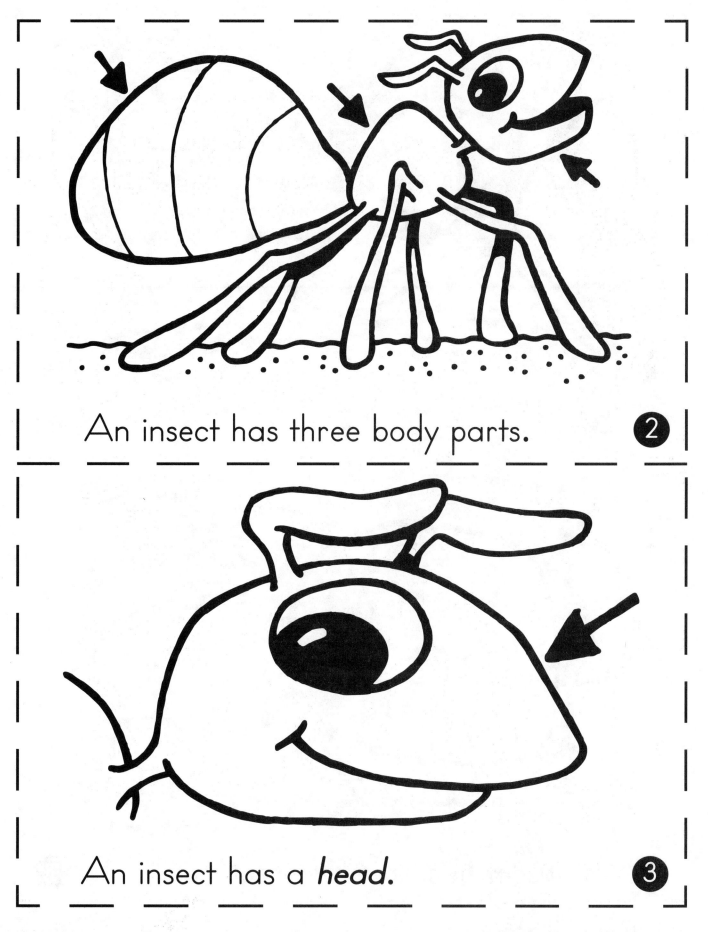

An insect has three body parts.   **2**

An insect has a *head.*   **3**

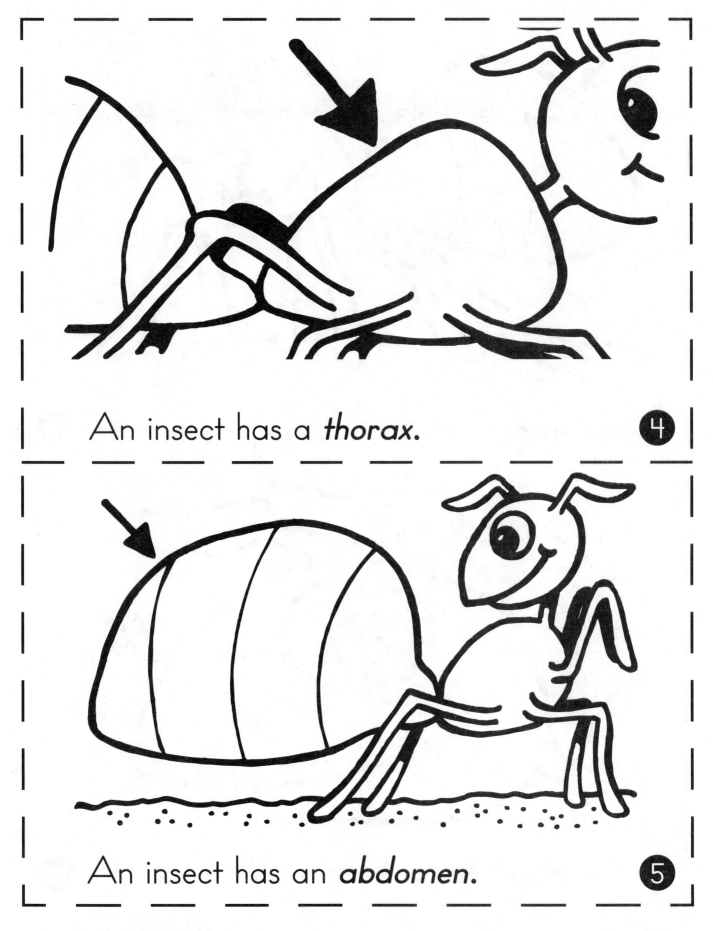

An insect has a *thorax.* **4**

An insect has an *abdomen.* **5**

Insect Life Cycle

By: _____

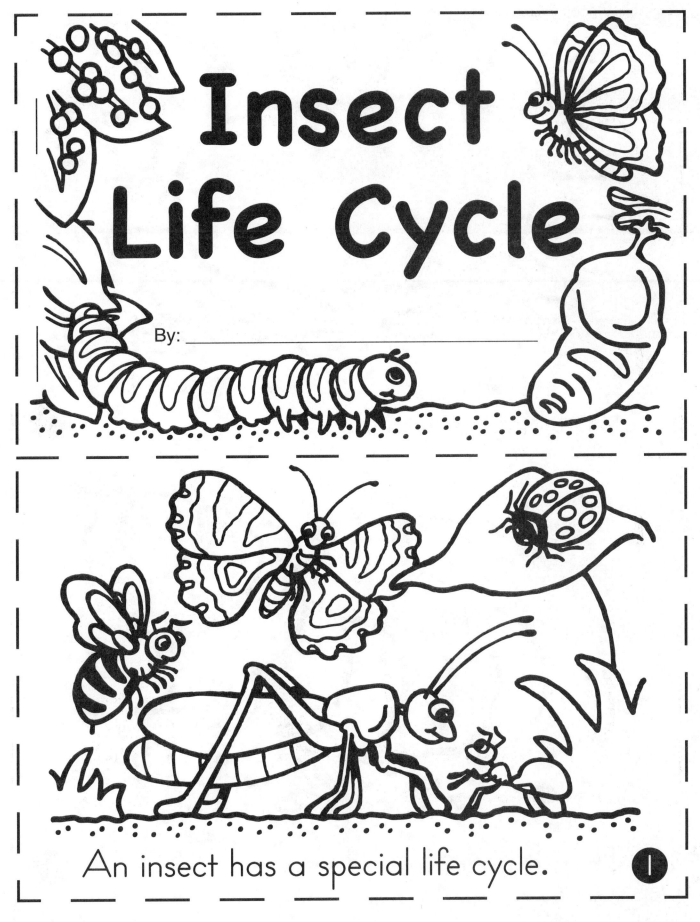

An insect has a special life cycle.

1

First it is an *egg*.

2

Then it becomes a *larva*.

3

Next it becomes a *pupa*. **4**

Last it is an adult *insect*. **5**

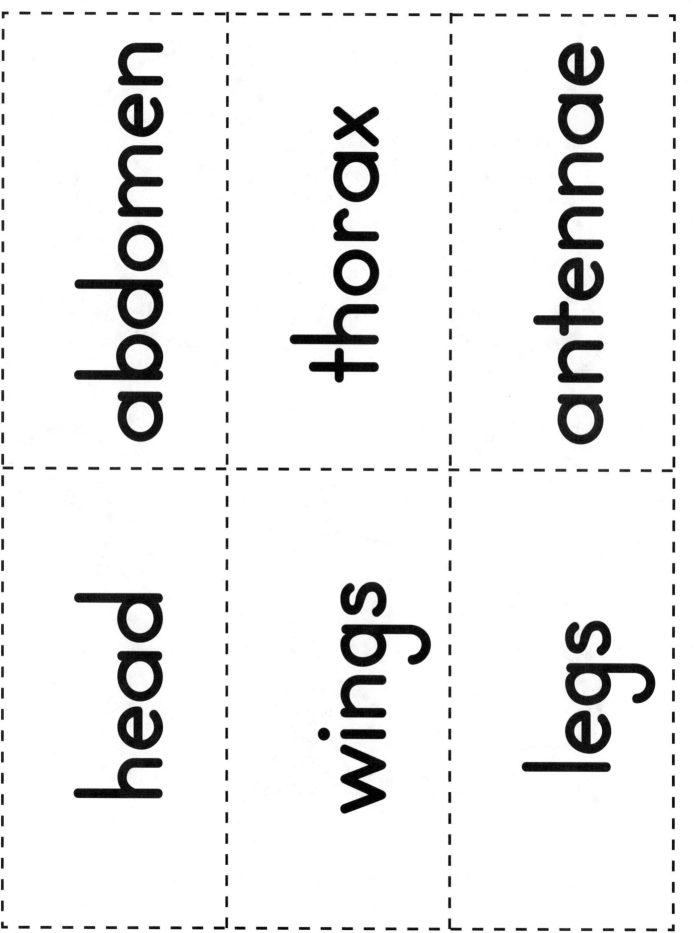

abdomen

thorax

antennae

head

wings

legs

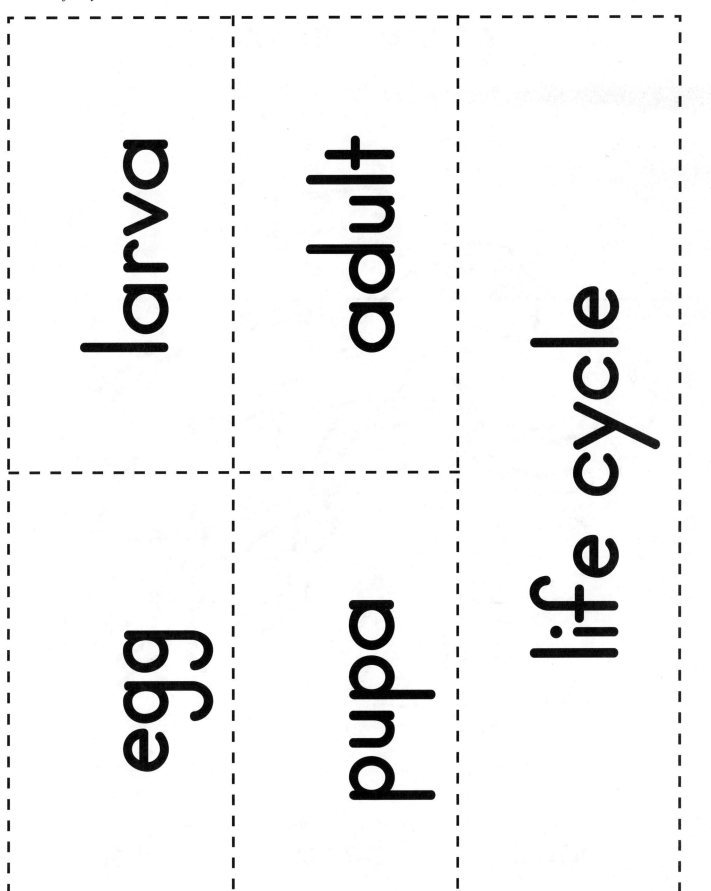

larva

adult

life cycle

egg

pupa

Parts of an Insect

Directions: Cut out and paste each word card on or by the correct part of an insect.

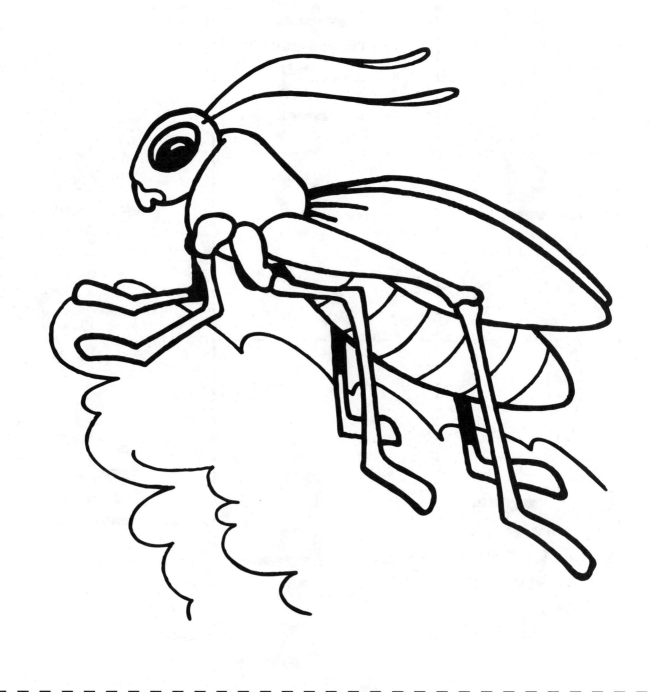

| thorax | head | legs |
| eye | abdomen | antennae |

Insect Life Cycle

Directions: Cut out and paste each word card by the correct life cycle stage of an insect.

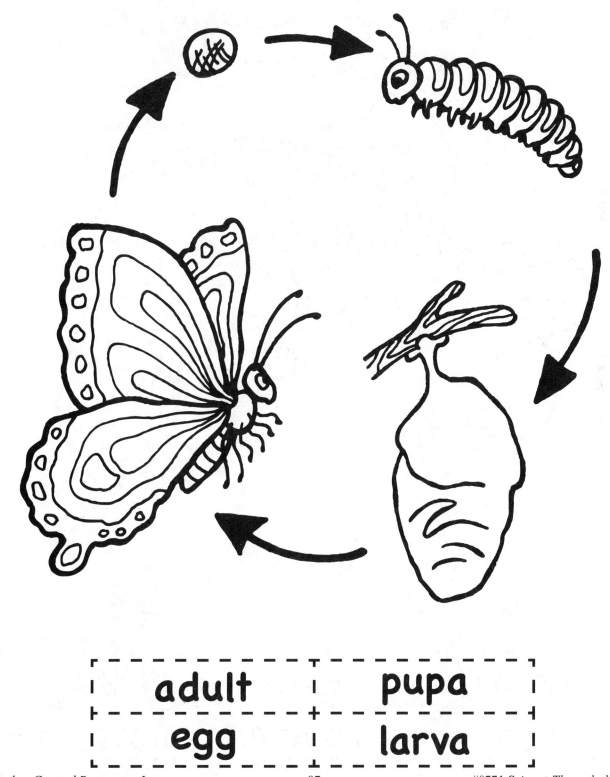

adult	pupa
egg	larva

Weather

lightning

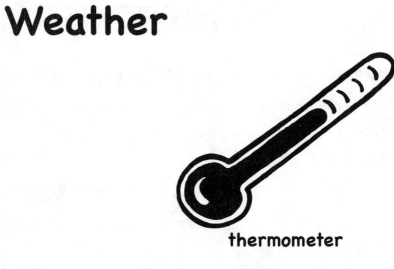

thermometer

rain

fog

sun

wind

snowflake

Weather Vocabulary

Lightning—bolt of electricity that forms during electrical storms

- thunder is the loud noise that happens when lightning strikes

Fog—low clouds that make it difficult to see

Wind—cold or warm air that moves (circulates)

Rain—water droplets that fall to the ground from clouds in the sky

- a rain gauge is an instrument for collecting rain water

Thermometer—measures air or water temperature in degrees

- a thermometer measures temperature in Centigrade or Fahrenheit degrees
- temperature is how hot or cold something is (e.g., the air)

Sun—the sun's rays heat up the air, land, and water

Snowflake—a six-sided crystal of snow

- tiny crystals fall and create a layer, which is called snow
- snow is water stored in the clouds that freezes as it falls to the ground

Seasons

Spring

Summer

Fall

Winter

Seasons Vocabulary

Note: The descriptions below reflect the Northern Hemisphere.

Spring—the mild season from March 21 to June 20
- new plants begin to grow and baby animals are born
- has lots of rain, as well as sun, so that new plants and animals can grow
- some fruits and vegetables ripen in spring (e.g., strawberries)
- days get longer (more sunlight) and temperature gets warmer

Summer—the sometimes harsh season from June 21 to September 20
- hot weather, along with possibility of rain and thunderstorms
- some fruits and vegetables ripen in summer (e.g., peaches)
- days are long (longest day is June 21)

Fall—the mild season from September 21 to December 20
- days get shorter (less sunlight) and temperature gets colder
- because there is less sunlight, the trees can begin to lose their leaves; leaves of some kinds of trees "fall" to the ground
- another name for fall is *autumn*
- some fruits and vegetables ripen in the fall (e.g., apples, pumpkins)

Winter—the sometimes harsh season from December 21 to March 20
- days get shorter and shorter (less sunlight) until December 21 (shortest day of the year)
- has snow, rain, sleet, hail, and windy conditions
- some trees lose all their leaves and are bare in winter; others are green all through winter (e.g., pine trees)

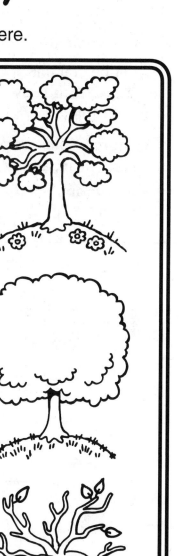

- -

Date _____

Dear Parent,

We will be starting a science unit on weather soon. There are a few items we need for the unit. Please provide the item circled below. If you are unable to provide the item, please let me know as soon as possible.

- unbreakable container
  (e.g., clean, empty plastic jar)
- dish soap
- small bottle of glycerin
  (available at the drugstore)
- several empty paper-towel tubes

- bubble wands (chenille stems can be bent and twisted into bubble wands)
- several old spoons or metal washers
- ball of brightly colored yarn
- wind sock or pinwheel (to borrow)
- wind chime (to borrow)

Please send your item to school with your child by _____.

Thank you for your help with this unit!

Sincerely,

- -

Date _____

Dear Parent,

We have just started a science unit on weather! We will be learning about different kinds of weather. Please talk about the weather with your child; for example, asking him or her what the weather is like today.

At the end of the unit, students will read *Weather* and *Seasons* Minibooks. Please ask your child to "read" his or her books to you and tell you about different kinds of weather and what types of weather happen during the seasons.

Sincerely,

- -

Weather

Background Information

Weather affects young children on a personal level. Weather determines how comfortable or uncomfortable the students will be on any given day. Knowing the weather forecast can help children decide what to wear to school. Weather determines what kinds of activities they can participate in (e.g., walking in the rain, swimming, playing in the snow). Not all children have had the same experiences with regard to weather. Some may not have seen snow firsthand; others may not know what the weather is like in a desert or near an ocean.

Unit Preparation

1. Purchase the following items at a grocery or discount store or ask for donations (see page 92):

   - small plastic containers (to collect rain water)
   - plastic beads
   - glitter glue
   - paper-towel tubes
   - metal washers or spoons
   - thermometers
   - tape
   - yarn

2. Copy and cut out the *Weather* and *Seasons* Word Cards (pages 120–122).

3. Copy and assemble the *Weather* and *Seasons* Minibooks (pages 113–119).

4. Copy and assemble the *Weather* Science Journals (pages 107–112).

5. Reproduce the *Weather* Assessment sheet (page 106).

6. Use chart paper or butcher paper to create the Inquiry Chart for Lesson 1 (see page 94), as well as the whole-class charts used throughout the unit.

Literature Links

A Drop of Water by Walter Wick

Anno's Counting Book by Mitsumaso Anno

Chicken Soup With Rice: A Book of Months by Maurice Sendak

Cloudy With a Chance of Meatballs by Judi Barrett

The Little House by Virginia Lee Burton

The Magic School Bus Kicks Up a Storm: A Book About Weather by Nancy White

The Mitten by Jan Brett

The Snowman by Raymond Briggs

The Snowy Day by Ezra Jack Keats

Weather by DK Eyewitness Books

Weather Words and What They Mean by Gail Gibbons

Weather (cont.)

Lesson 1: Introducing the Weather Unit

Tell the children that the class is going to study the weather, and that they all know a lot about the weather because they live with it every day! Ask each student to turn to a neighbor and tell him or her something about the weather. Allow a few students to share their examples with the class. Have the children help you complete the Weather Inquiry Chart (see below). (Whole Class Assessment)

Note: Color-code each column of the chart using bright markers such as red, blue, and green.

Weather Inquiry Chart

What Do You Know About Weather?	What Do You Want to Learn About Weather?	How Can We Find Out?

The inquiry chart serves as an assessment. It demonstrates what the children already know, what their misconceptions may be, and what they are wondering about. It is an excellent springboard for future lessons. Inquiry charts can provide teachers with needed information when planning lessons.

The Sample Weather Inquiry Chart (below) gives an idea of the kinds of responses young children might give.

Sample Weather Inquiry Chart

What Do You Know About Weather?	What Do You Want to Learn About Weather?	How Can We Find Out?
There's snow, rain, thunderstorms, snowstorms, and sunlight. Weather can be bad or good. There are different kinds of weather.	How does the sky get the water from the ground? How do the clouds make the rain go down? Where do clouds come from?	Read books. Look in an encyclopedia. Look on the Internet. Ask a teacher. Study the weather.

Note: At the end of each unit, you can have the children help you complete a chart of "What We Learned" as a whole group culminating assessment.

Weather *(cont.)*

Lesson 2: Weather Picture Chart

1. Gather the children together on the rug near the *Weather* Picture Chart (see page 88) to teach content and vocabulary. Show the children the chart paper with the pre-drawn pencil diagram of types of weather. (Be sure to write your teacher notes lightly in pencil next to each part of the diagram.)

2. Talk about weather in "chunks," providing key vocabulary (*lightning, fog, wind, rain, thermometer, sun, snowflake*) and content (*weather is related to seasons, the sun warms the earth*). As you speak, trace over the corresponding parts with a permanent marker. Do not label the diagram yet.

   Teacher Note: Describe what you are doing as you draw. Mention the curving lines, larger and smaller shapes, etc. Your descriptions should help students feel more comfortable with their own Science Journal illustrations.

3. After tracing over the entire picture and providing the vocabulary and content, ask the children to tell what they learned about the weather.

4. Write the children's responses on appropriate places on the chart (i.e., labeling types of weather).

5. Introduce the *Weather* Word Cards (pages 120–121) and add them to a pocket chart.

6. Display the completed *Weather* Picture Chart in the classroom for a reference tool.

 Weather Home Link: Have the children complete the *Types of Weather* worksheet (page 123) for homework.

Lesson 3: Class Weather Chart

Each day during calendar time, ask the children what the weather is like. Have a student attach a picture to the calendar of either windy, sunny, rainy, cloudy, or snowy. The student may attach the question mark picture to the calendar if the weather that day fits into more than one category. (Use the Weather Chart Pictures on page 96.)

Inquiry Skill—Observation

Weather Chart Pictures

Weather *(cont.)*

Lesson 4: Literature Connection

1. Introduce the story *Cloudy With a Chance of Meatballs* by Judi Barrett to the whole class. Talk about the front cover of the book, the author, and illustrator.

2. Do a "picture walk" first. Show the children each page of the book without reading the text and ask them to predict what will happen at various points in the story.

3. Read the story to the class. Ask the children to share what they liked best about the story.

4. Have the children help you complete a chart that lists the strange types of weather described in the book. Have the children help you complete a Plot Analysis and Plot Elements Chart (see samples below). (Whole Class Assessment)

Sample Plot Analysis Chart

Strange Weather in *Cloudy With a Chance of Meatballs*	
• Raining soup	• Jell-O® sunset
• Snowing mashed potatoes and peas	• Pearl soup fog
• Windstorm of hamburgers	• Salt and pepper wind
• Brief shower of orange juice	• Tomato tornado
• Mustard clouds	• Bad storm of giant meatballs
• Drizzle of soda	

Sample Plot Elements Chart

Beginning	Middle	End
Grandpa is making breakfast.	Grandpa tells the story of the town of Chewandswallow and how the weather is actually food! The weather gets really bad.	The children go outside to play in the (real) snow and imagine mashed potatoes.

Weather *(cont.)*

Lesson 5: Rain Activity

1. Children bring in various sized containers from home to make homemade rain gauges. Clear containers (e.g., empty, plastic jars) with labels removed work well for this activity. Help the children mark inch measurements on the sides of the containers with a permanent marker and/or masking tape.

2. On a rainy day, place the rain gauges outside the classroom.

3. After one hour, bring in the rain gauges and have the children record the amount of rain collected in their *Weather* Science Journals (Observation 1, page 108).

4. Place the rain gauges outside when it rains again. After an hour, bring in the rain gauges again and have the children record the total amount of rain collected in their *Weather* Science Journals (Observation 1, page 108).

Inquiry Skills—Observation, Compare and Contrast, Measurement, Recording Data

Lesson 6: Snow Activity

Materials

- 10 oz. (295.7 mL) glass
- thermometer
- snow or crushed ice

Directions

1. If you live in an area where it snows, take the children outside on a snowy day and collect enough snow to fill the glass. If there is no snow available, use crushed ice.

2. Place the thermometer in the glass and read the temperature to the class.

3. Take the class inside and have them sit on the rug. Place the glass where all the students can see it.

4. It should take about 5 minutes to melt the snow/crushed ice. While it is melting, talk to the class about the changes that occur when substances melt (*solid changes to liquid*).

5. After the snow has melted, read the temperature on the thermometer again. Compare the first temperature reading to the second temperature reading. Discuss with the class.

6. Have the students record the results in their *Weather* Science Journals (Observation 2, page 109).

Inquiry Skills—Observation, Measurement, Compare and Contrast

Weather *(cont.)*

Lesson 7: Making Snowflakes

Materials

- *A Drop of Water* by Walter Wick (or another book that shows magnified snowflakes)
- scissors
- 8" x 8" (20.3 cm x 20.3 cm) squares of white paper, one per student

Directions

1. Ask the children if they have ever seen a snowflake up close. What do they look like?

2. Show the children pictures of real snowflakes magnified many times in the book. Talk about how each snowflake is unique—no two are alike. Talk about how when tiny water droplets freeze as they fall from the sky, they sometimes become snowflakes.

3. Allow the children time to discuss the pictures of the snowflakes and ask you questions. Ask them to tell a neighbor which snowflake picture is their favorite and why.

4. Show the children how to fold white paper and cut out small circles and triangles to make their own paper snowflake.

5. Guide the children through making their own paper snowflakes. Use the directions below and the illustrations on page 100.

Make a Snowflake

a. Fold the paper in half.

b. Fold the paper in half again.

c. Fold the paper in half diagonally to form a cone.

d. Cut off the tip of the cone.

e. Cut off the top of the cone, either a round cut or a straight cut.

f. Cut small triangles and circles out of the sides and top of the cone.

g. Carefully unfold the paper to reveal the snowflake.

Inquiry Skills—Observation, Compare and Contrast

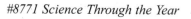

Sample Snowflake

Weather *(cont.)*

Lesson 8: Wind Activity

Materials

- plastic cup (for each student)
- bubble wand (for each student)
- paper towel (for each student)
- plastic tub to carry materials
- 8 1/2" x 11" (21.6 cm x 27.9 cm)
- directional cards (N, S, E, W)

Bubble Solution

- 8 cups (2.4 L) water
- ½ cup (118.3 mL) liquid dish soap (Dawn®)
- 3 teaspoons (14.8 mL) glycerin

Combine ingredients in a plastic bottle and swirl, don't shake, to mix.

Note: Bubble solutions can be purchased.

Directions

1. Make directional cards. Place the cards outside. Weigh the directional cards down with rocks.
2. Explain to the children that they will go outside and blow bubbles to see which direction the wind is blowing.
3. Distribute a cup, a paper towel, and a bubble wand to each student.
4. Pour 3 ounces (88.7 mL) of bubble solution into each child's cup.
5. Have the children stand in three different places on the playground and blow bubbles. Instruct them to observe which way the wind is blowing and record their observations in their *Weather Science Journals* (Observation 3, page 110).

Inquiry Skill—Observation

Lesson 9: Wind Chimes

Materials (for each student)

- paper-towel tube
- 5 plastic beads
- markers
- glitter glue (optional)
- 4 (old) metal spoons or metal washers
- 4 pieces of yarn, cut into 12" (30.5 cm) lengths
- piece of yarn, cut into a 24" (70 cm) length

Directions

1. An adult should carefully prepare the paper-towel tube in advance so that it measures 6" (15.3 cm). Use a toothpick or paper clip to poke four holes, 1" (2.5 cm) apart, along one side of the tube.
2. Tie a knot in each of the lengths of yarn. Thread the yarn through each of the holes so that the knot remains on the inside of the tube.
3. Decorate the tube with markers and glitter glue. Be creative and use all the available space. Let dry.
4. Thread a bead onto each length of yarn.
5. Tie a spoon or washer onto the end of each length of yarn using a double knot.
6. Thread the 24" (70 cm) length of yarn through the hole in the tube, leaving the ends of yarn hanging out of each end of the tube. Bring the two ends of yarn together and thread a bead through both ends. Tie to secure.
7. Direct the children to take the wind chimes outside and hold them up in the air with their hands to see what happens when the wind blows.

Inquiry Skill—Observation

Weather *(cont.)*

Lesson 10: Sun Activity

Materials

- thermometer

Directions

1. Explain to the children that they will be working in small groups to measure the temperature in three locations on the playground: in full shade, full sunlight, and partial sun/shade.

2. Give each group a thermometer and take the children outside and have them find their first location, read the temperature on their thermometer, and record the measurement in their *Weather Science Journals* (Observation 3, page 109).

3. Groups repeat step 2 twice, using two other locations.

4. Take the children back inside the classroom. Hold a class discussion about the differences in temperature they found in full sunlight, partial sun/shade, and full shade. Record the results.
Inquiry Skills—Observation, Measurement, Recording Data

Sample Playground Temperature Graph

Group	Full Sunlight	Partial Sun/Shade	Full Shade
Patrick, Katie, Alex, Maher	80 degrees	77 degrees	72 degrees
Kerstie, Madison, Ben, Tony	79 degrees	77 degrees	72 degrees
Lonnie, Michael, Hannah, Maria	80 degrees	76 degrees	72 degrees

Lesson 11: Rainbow Activity

Materials (for each small group)

- 8 oz. (236.6 mL) clear, plastic cup
- water

- white paper
- crayons in the following colors: red, orange, yellow, green, blue, and purple

Directions

1. Explain to the children that they will be going outside to make their own rainbow using sunlight and a cup of water. Distribute materials to groups.

2. Take the children outside and have them find a place in the sun for their group to work.

3. Have the children fill their plastic cup three-fourths full with water, set the piece of paper on the blacktop, and then carefully set the cup on the paper. Students take turns carefully lifting up the water cup and tilting it gently so that the sun's rays shine through the water and onto the paper in the form of color spots.

4. Students record the colors they see in their *Weather* Science Journals (Observation 5, page 112).

Inquiry Skills—Observation, Recording Data

Weather (cont.)

Lesson 12: Weather Minibook and Big Book

1. Distribute a *Weather* Minibook (pages 113–116) to each child.

2. Read the book as a class. Discuss the illustrations. Point out key vocabulary (*rainy, windy, sunny, cloudy, snowy*). Review the *Weather* Word Cards (pages 120–121).

3. Explain what the children are to do at each center.

   Center 1: Read the *Weather* Minibook with a partner. Take turns retelling the sequence of events.

   Center 2: Color the pictures in the *Weather* Minibook.

   Center 3: Use a yellow crayon to highlight key vocabulary: *rainy, windy, sunny, cloudy,* and *snowy.* Using a red crayon, circle the high frequency word *are.* Draw a box around the word *weather* each time you see it. (Assessment)

   Center 4: Listen to a taped version of *Cloudy With a Chance of Meatballs* or the *Weather* Minibook.

4. At the end of the unit, each child will take home the *Weather* Minibook and "read" it to his or her family.

5. Create an additional class book by enlarging the *Weather* Minibook. Students can take turns coloring the pages. Place the completed *Weather* Big Book in the classroom library.

Lesson 13: Seasons Picture Chart

1. Gather the children together on the rug near the *Seasons* Picture Chart (see page 90) to teach content and vocabulary. Show the children the chart paper with the pre-drawn pencil diagram of the seasons. (Be sure to write your notes lightly in pencil next to each part of the diagram.)

2. Talk about the seasons in "chunks," providing key vocabulary (*spring, summer, fall, winter*) and content (*in spring, rain helps new plants grow; summer has hot weather and days are longer; in fall, days get shorter and trees lose leaves; winter has snow and days are shorter*). As you speak, trace over the corresponding parts with a permanent marker.

3. After tracing over the entire picture and providing the vocabulary and content, ask the children to tell what they learned about the seasons.

4. Write the children's responses on appropriate places on the chart.

5. Introduce the *Seasons* Word Cards (page 122) and add them to a pocket chart.

6. Display the completed *Seasons* Picture Chart in the classroom for a reference tool.

 Seasons **Home Link:** Have the children complete the *Seasons and Weather* worksheet (page 124) for homework.

Lesson 14: Weather and Seasons

Hold a class discussion. Talk about which kinds of weather we usually have during each season (e.g., sun and heat in summer, cool and wind in fall, snow and cold in winter, rain and fog in spring). Have the students help create a Weather and Seasons Chart, describing what kind of weather goes with each season.

Inquiry Skills—Observation, Compare and Contrast

Weather *(cont.)*

Lesson 15: What to Wear?

Hold a class discussion. Ask the student to help create a chart that tells what to wear for each type of weather (see below). Suggest that they cut out articles of clothing from magazines and catalogs.

Sample Chart: What Should We Wear?

Rainy	Windy	Snowy	Sunny	Cloudy
rain coat	jacket	coat	cap	long sleeved shirt
rain boots	long pants	mittens	shorts	long pants
umbrella	long sleeved shirt	warm hat	T–shirt	tennis shoes

Lesson 16: *Seasons* Minibook and Big Book

1. Distribute a *Seasons* Minibook (pages 117–119) to each child.

2. Read the book as a class. Discuss the illustrations. Point out key vocabulary (*spring, summer, fall, winter*). Review the *Seasons* Word Cards (page 122).

3. Explain what the children are to do at each center.

   Center 1: Read the *Seasons* Minibook in small groups with the teacher.

   Center 2: Fill in words for each page of the *Seasons* Minibook.

   Center 3: Draw a picture for each page of the *Seasons* Minibook that matches the word you wrote. (Assessment)

   Center 4: Listen to a taped version of *Chicken Soup With Rice* or the *Seasons* Minibook.

4. At the end of the unit, each child will take home the *Seasons* Minibook and "read" it to his or her family. Encourage children to share what they have learned about seasons with their families.

5. Create an additional class book by enlarging the *Seasons* Minibook. Students can take turns coloring the pages. Place the completed *Seasons* Big Book in the classroom library.

Weather *(cont.)*

Culminating Activity: What Can I Do Outside?

Lead a class discussion about what kinds of activities we do during certain kinds of weather and certain seasons. Create a class chart that lists the activities (see sample below).

Sample Weather Activities Chart

Summer	Fall	Winter	Spring
swimming go to beach	play in leaves football	play in snow skiing	jump in puddles baseball

Assessment Pages: Demonstrating Content Knowledge and Vocabulary

- **Weather Bears** (page 106)—Show which type of weather each bear is dressed appropriately for.

- **Types of Weather** (page 123)—Match each *Types of Weather* word (*sunny, rainy, cloudy, windy, snowy, foggy*) to the corresponding picture.

- **Seasons and Weather** (page 124)—Match each type of weather (*snow, rain, wind, clouds, fog, thunder, lightning*) to the appropriate season. Draw pictures and label them.

Weather *(cont.)*

Culminating Assessment: Weather Bears

Objective: The students will demonstrate they understand that different clothing is appropriate for different types of weather by matching an outfit to the appropriate type of weather.

Directions: Cut out the bears. Glue each bear under the matching type of weather.

Rainy   ### Sunny

Snowy   ### Cloudy

Science Journal

WEATHER

By: _____

Science Journal

Observation 1: Rain

1. Draw a picture of your rain gauge.

2. Mark the level of the water after the first hour.

3. Mark the level of the water after the second hour.

Describe: How does a rain gauge help us measure weather?

_____

_____

_____

_____

Word Bank

weather   rain   collect   amount

Science Journal

Observation 2: Snow

1. On the thermometer, mark the first temperature of the glass of snow.

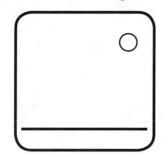

2. On the thermometer, mark the temperature of the snow after it melts.

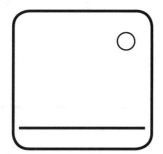

F°		C°
110		40
100		30
90		
80		30
70		20
60		
50		10
40		
30		0
20		
10		-10
0		-20

Describe: How does a thermometer help us measure weather?

_____

_____

_____

_____

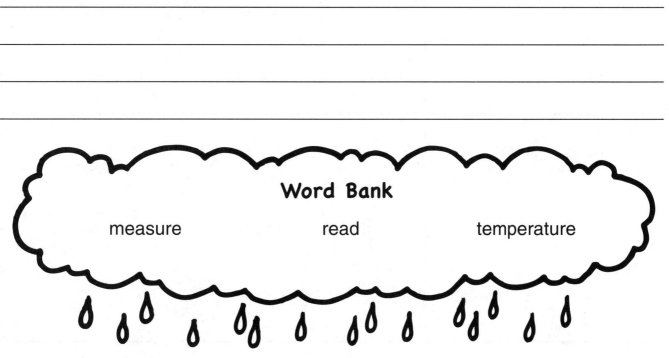

Word Bank

measure   read   temperature

Science Journal

Observation 3: Wind

1. Blow bubbles in three locations.

2. Watch to see which direction your bubbles go.

3. Record your results on the chart. Circle one direction for each location.

Wind Direction

Location 1	N S E W
Location 2	N S E W
Location 3	N S E W

Science Journal

Observation 4: Measuring Temperature

1. Draw a picture of each location.

2. Write a description of each location.

3. Record the temperature for each location.

	Picture	Location	Temperature
Full Shade		_____ _____ _____	
Full Sun		_____ _____ _____	
Partial Sun and Shade		_____ _____ _____	

Observation 5: Rainbow

1. Gently tilt the cup of water from side to side.

2. Record the colors you see on the paper.

Weather

By: _____

There are different kinds of weather.

①

Some days are *rainy.*   **2**

Some days are *sunny.*   **3**

Some days are *windy*. **4**

Some days are *cloudy*. **5**

Some days are *snowy*.　　　**6**

My favorite weather is _____.　　**7**

Seasons

By: _____

My favorite thing to do in the summer is _____.
The weather is _____.

①

My favorite thing to do in the fall
is _____ .

The weather is _____ . **2**

My favorite thing to do in the winter
is _____ .

The weather is _____ . **3**

My favorite thing to do in the spring
is _____ .

The weather is _____ . ❹

My favorite season is _____
because _____ . ❺

rain

wind

thermometer

fog

sun

lightning

snowflake

temperature

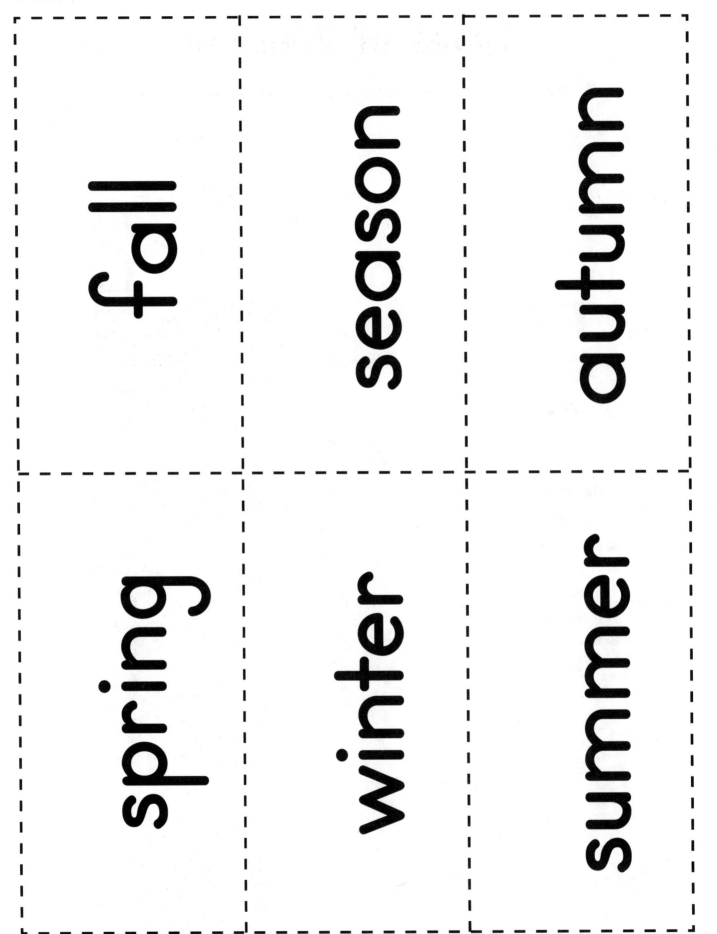

fall

season

autumn

spring

winter

summer

Types of Weather

Directions: Draw a line to match each word to the correct picture.

sunny

rainy

cloudy

windy

snowy

foggy

Seasons and Weather

Directions: Draw weather for each season. Label using words from the Word Bank.

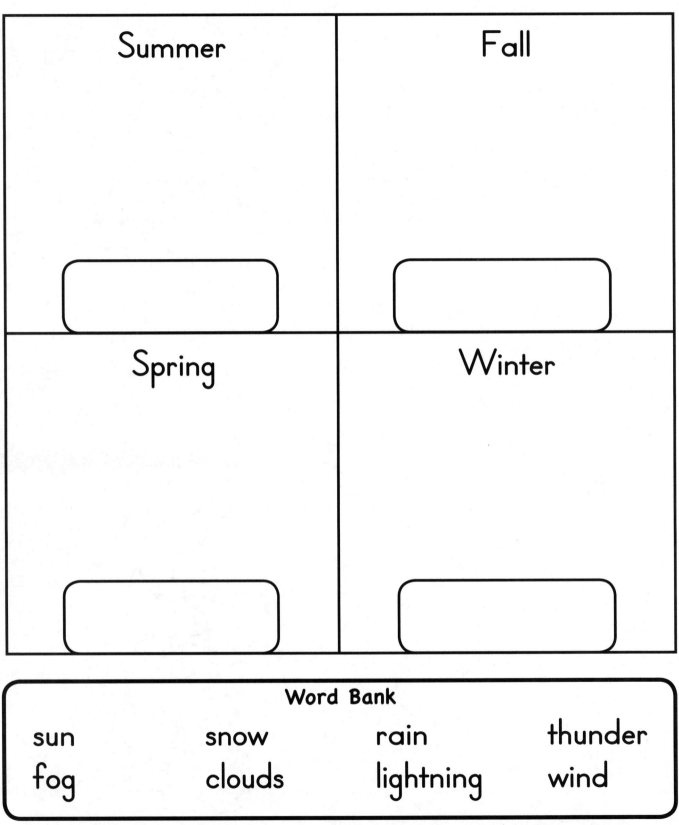

Summer

Fall

Spring

Winter

Word Bank

sun	snow	rain	thunder
fog	clouds	lightning	wind

States of Matter

Liquids

Gases

Solids

States of Matter Vocabulary

Matter—everything with mass that is found in the universe

States—the different forms matter exists in

Liquid—matter that can flow
- settles to the bottom of a container
- takes the shape of the container it is in
- water is an example of a liquid

Solid—matter that keeps a constant shape
- loses shape if heated enough
- ice is an example of a solid

Gas—matter that has no shape of its own
- expands to fill the space available
- mostly colorless and transparent
- steam is an example of a gas

- -

Date _____

Dear Parent,

We will be starting a unit on matter soon. Matter makes up everything in the universe and comes in different forms: liquid, solid, and gas. Please provide the item circled below. If you are unable to provide the item, please let me know as soon as possible.

- box of quart-sized resealable plastic bags
- box of gallon-sized resealable plastic bags
- gallon of homogenized milk
- vanilla extract
- 1 cup granulated sugar
- box of plastic spoons
- bag of toy balloons

- box of wrapped drinking straws
- box of Borax®
- gallon of distilled water
- box of food coloring —4 colors
- roll of crepe paper
- box of cornstarch
- box of rock salt

Please send your item to school with your child by _____.

Thank you for your help with this unit!

Sincerely,

- -

Date _____

Dear Parent,

We have just started a science unit on matter! We will be working with liquids, solids, and gases to discover the states and properties of matter. Please ask your child what he or she is learning about in science over the next few weeks. Play a game with your child by naming something (e.g., air, toy, juice) and having the child tell you what type of matter it is (e.g., gas, solid, liquid).

At the end of the unit, students will make homemade ice cream in class and will take home *States of Matter* and *Changes in States of Matter* Minibooks they have been reading in class. Please ask your child to "read" his or her books to you and tell you what he or she knows about the different states of matter.

Sincerely,

- -

- -

Date _____

Dear Parent,

We are now wrapping up our science unit on matter (liquids, solids, gases). Attached are some simple recipes that allow children to see changes in states of matter. These recipes are fun to make with your children and are tasty, too! Please take some time and try these out with your family.

As you and your child make these recipes, talk about the changes that occur. For example, the coldness of the freezer allows juice to freeze into popsicles and the boiling heat of the stove and the cooling afterward makes the gelatin harden into a solid.

Enjoy!

Sincerely,

- -

Recipe Cards

Frozen Water Bottle

Objective: A parent supervises his or her child to change a liquid to a solid by removing heat.

Material
- small plastic water bottle (unopened)

Directions

1. Observe the contents of a water bottle without opening it. Ask your child, *Is the water a liquid, solid, or gas?*

   Answer: liquid

2. Place the water bottle in the freezer.

3. After 1–3 hours, check the water bottle. Ask your child, *What has happened to the water?*

   Answer: The water is partially frozen.

4. After 4–8 hours, check the water bottle again. Ask your child, *What has happened to the water?*

   Answer: Water, a liquid, froze into ice, a solid.

- -

Recipe Cards *(cont.)*

Chocolate Chip Cookies

Objective: A parent supervises his or her child to mix liquid and solid ingredients and heat them.

Materials

- mixing spoon
- small mixing bowl
- large mixing bowl
- teaspoon
- ungreased cookie sheets
- spatula
- wax paper or cooling rack

Liquid Ingredients

- 1 cup (237 mL) butter, softened until liquid
- 2 eggs
- 1 teaspoon (5 mL) vanilla

Solid Ingredients

- 2¼ (532 mL) cups flour
- ¾ (177 mL) cup granulated sugar
- ¾ (177 mL) cup brown sugar
- 1 teaspoon (5 mL) salt
- 1 teaspoon (5 mL) baking soda
- 2 cups (473 mL) chocolate chips

Directions

1. Preheat oven to 375°F (190.6° Celsius).
2. Combine all liquid ingredients in the small mixing bowl. Stir with a spoon to mix. Ask your child, *Is this mixture a liquid or a solid?* **Answer:** liquid
3. Combine all solid ingredients in the large mixing bowl. Stir with a spoon to mix. Ask your child, *Is this mixture a liquid or a solid?* **Answer:** solid
4. A little at a time, add the liquid mixture to the solid mixture. Stir after each addition until you finished combining the ingredients. Ask your child, *Is this mixture a solid or a liquid?* **Answer:** solid—but pliable like play dough
5. Drop by teaspoonfuls onto an ungreased cookie sheet.
6. Bake at 375° for 12 minutes.
7. Remove cookie sheets from oven. Let cool 5 minutes then use a spatula to remove cookies from tray. Cool cookies for 10 minutes on a sheet of wax paper or a cookie rack.
8. Serve cookies. Ask your child, *Are these cookies liquid or solid?* **Answer:** solid—hard and not pliable. *What made the cookies become solid?* **Answer:** heat

Recipe Cards (cont.)

Gelatin

Objective: A parent supervises his or her child to mix liquid and solid ingredients, heat the mixture, and cool it into a solid.

Safety Note: Children must be supervised at all times near the stove. Parents must boil and pour the water.

Materials
- pan (for boiling water)
- glass bowl

Solid Ingredient
- packet gelatin dessert (any flavor)

Liquid Ingredient
- water (as directed on gelatin package)

Directions
1. Boil the water.
2. Pour boiling water into a glass bowl. Add contents of gelatin packet. Stir to mix thoroughly. Ask your child, *Is this mixture a liquid or a solid?* **Answer:** liquid
3. Cool the mixture in refrigerator until set.
4. When completely set, serve the gelatin. Ask your child, *What happened to the liquid?* **Answer:** The heat and then the cold caused it to become a solid.

Juice Pops

Objective: A parent supervises his or her child to change a liquid to a solid by removing heat.

Materials
- 5 three-ounce (88.7 mL) paper cups
- 5 wooden craft sticks
- aluminum foil

Liquid Ingredient
- 2 cups (473 mL) apple, orange, or grape juice

Directions
1. Carefully divide the juice among the five paper cups.
2. Cover each cup with aluminum foil.
3. Insert one craft stick through the center of each foil cover.
4. Place the cups in the freezer for three hours or until set.
5. To remove the juice pops, carefully run warm water over side of cup to loosen. Ask your child, *What happened to the liquid?* **Answer:** It became a solid. *How did it change to a solid?* **Answer:** by taking away heat

Matter

 ## Background Information

All materials on Earth are made of matter. Matter exists in different states. Primary grade children study the following states: liquid, solid, and gas. Each state has distinct properties that can be observed and described. *Liquids* flow and take the shape of the container they are in. *Solids* have a constant shape. *Gases* expand to fill the space available and are usually invisible. When matter is mixed, heated, or cooled, the properties and states can change.

Unit Preparation

1. Purchase the following items at a grocery or discount store or ask for donations (see letter on page 127):

   - box of quart-sized resealable plastic bags
   - box of gallon-sized resealable plastic bags
   - gallon of homogenized milk
   - vanilla extract
   - 1 cup granulated sugar
   - plastic spoons, cups, and containers
   - bag of toy balloons
   - box of wrapped plastic drinking straws
   - box of Borax®
   - tea kettle

   - gallon of distilled water
   - box of food coloring (including green)
   - box of cornstarch
   - rock salt
   - ice
   - sample liquids and solids
   - white glue
   - containers of play dough
   - roll of crépe paper
   - hot plate

2. Copy and cut out the *Matter* Word Cards (page 170).

3. Copy and assemble the *States of Matter* and *Changes in States of Matter* Minibooks (pages 164–169) and the *Matter* Science Journals (pages 157–163).

4. Reproduce the *Matter* Assessment sheets (pages 153–156).

5. Use chart paper or butcher paper to create the whole-class charts used throughout the unit.

6. Prepare materials for the Mixing Liquids Investigations (see pages 136, 138, 140, 142): Write *Mystery Liquid #1* on a plastic container. Add 2 tablespoons Borax® to 2 cups distilled water and stir. Write *Mystery Liquid #2* on a plastic container. Add 2 cups distilled water only.

7. Reproduce one Rebus Investigation Sheet (see pages 137, 139, 141, 143) and a corresponding Group Response Sheet (see pages 144–147) for each small group of four students.

8. Cut sheets of 9" x 12" (22.9 x 30.5 cm) construction paper in half to form two 4 ½" x 12" (11.4 x 30.5 cm) strips. Tape these together to form a 4 ½" x 24" (11.4 x 61 cm) strip for each child to use for the Culminating Assessment (see page 152).

Literature Links

Bartholomew and the Oobleck by Dr. Seuss

Solids, Liquids, and Gases by Louise Osborne

What Is the World Made Of? All About Liquids, Solids, and Gases by Kathleen Weidner Zoehfeld

Matter *(cont.)*

Lesson 1: Introducing the Matter Unit

Ask the children if they have ever participated in the following activities.

- played with blocks

- frozen water to make ice

- played with sand or water at the beach

- watched steam come from a tea kettle

If they have, then they know about matter! Tell the children that matter makes up all the things around them. Tell them that they are starting a science unit all about matter.

Lesson 2: *States of Matter* Picture Chart

1. Gather the children together on the rug near the *States of Matter* Picture Chart (see page 125) to teach content and vocabulary. Show the children the chart paper with the pre-drawn pencil diagram of states of matter. (Be sure to write your teacher notes lightly in pencil next to each part of the diagram.)

2. Talk about matter in "chunks," providing key vocabulary (*states, liquids, solids, gases*) and content (*liquids flow and take the shape of the container they are in, solids are hard, gases can sometimes be seen and are sometimes invisible*). As you speak, trace over the corresponding parts with a permanent marker. Do not label the diagram yet.

   Teacher Note: Describe what you are doing as you draw. Mention the curving lines, larger and smaller shapes, etc. Your descriptions should help students feel more comfortable with their own Science Journal illustrations.

3. After tracing over the entire picture and providing the vocabulary and content, ask the children to tell what they learned about the states of matter.

4. Write the children's responses on appropriate places on the chart (i.e., labeling states of matter).

5. Introduce the *Matter* Word Cards (page 170) and add them to a pocket chart.

6. Display the completed *States of Matter* Picture Chart in the classroom for a reference tool.

 Matter **Home Links:** Have the children complete the *States of Matter* worksheet (page 171) and *Changes in States of Matter* (page 172) worksheet for homework.

Matter *(cont.)*

Lesson 3: Playing with Matter

Materials (for each small group of four students)

- 5–6 wooden blocks
- gallon-sized resealable plastic bag containing a thick liquid (e.g., syrup, shampoo)
- gallon-sized resealable plastic bag containing a thin liquid (preferably water)
- 5–6 seashells
- 3 metal bolts and 3 metal screws

- 3 gallon-sized resealable plastic bags (filled with air and sealed)
- balloon (inflated with air and tied)
- chart paper divided into 7 columns and 2 rows
- markers
- timer

Safety Note: Before the children begin to explore the materials, remind them of safety and behavior expectations (i.e., do not open the bags, do not hit the balloon to each other, do not throw any materials).

1. Distribute materials to Materials Managers and set the timer for 5 minutes. Let children freely explore the materials.

2. After the timer goes off, get the children's attention. Show them a Sample Properties of Matter Chart (see below). Ask the Reporter in each group to give a few examples of properties he or she noticed about the materials. Demonstrate how to record responses on the chart.

3. Set the timer for 10 minutes. Invite groups to continue exploring the materials. Have them describe what the materials look like (e.g., shiny) and how the materials feel.

4. Have the Recorders write titles for each column (copy class chart) and words that describe the properties of each type of matter in each column of the group chart. All group members may illustrate the words.

5. After the groups have had a chance to complete their charts, ask the Recorders to share the ideas to help complete the class chart. Record each group's responses.

6. When the chart is complete, lead the children in a discussion of the different types of matter. Provide the children with the vocabulary: *liquids, solids, gases*. Write the words on the board.

7. Discuss examples of each type of matter. Have the groups identify each type of matter on the chart. The group can devise the way, either by color-coding (circling titles) or labeling.

8. Have the children use their *Matter* Science Journals to draw a picture of one example of matter. Then, instruct the students to write 2–3 sentences describing the matter (Observation 1, page 158). (Assessment)

Inquiry Skills—Observation, Recording data

Sample Properties of Matter Chart

Wooden Blocks	Syrup	Water	Air in balloon	Seashells	Air in Bag	Bolts
hard smooth	flows slowly	pour it moves around flows fast	round shaped	hard breakable	invisible shaped like the bag	shiny hard can't break

Matter (cont.)

Lesson 4: States of Matter

1. Ask the children to help you complete a States of Matter Class Chart (see below), based on the exploration they did with matter in Lesson 3: Playing with Matter (see page 133).

2. Display this chart in the classroom as it will be added to in later lessons.

3. Instruct the students to copy words from the class chart to their individual charts in their *Matter* Science Journals (States of Matter Chart, page 159).

Inquiry Skills—Observation, Recording Data

Sample States of Matter Class Chart

Liquids	Solids	Gases
flow	solid or hard	can be visible float in air

Note: Lessons 5–8 provide example demonstrations that can be done as a whole class. The teacher should record student responses in the appropriate column of the States of Matter Class Chart (see sample above).

Lesson 5: Liquids Demonstration/Activity

Materials

- 4 plastic cups
- water
- 2–3 wooden blocks
- quart-sized resealable plastic bag (containing water)
- quart-sized resealable plastic bag (containing syrup or shampoo)

Directions

1. Pour water from one cup to another. Ask the students, "Is this a liquid? How do you know?"

2. Provide another example for the students: Pour a cup of wooden blocks into another cup. Ask the students, "Is this a liquid, gas, or solid? How do you know?"

3. Compare two liquids in the two plastic bags (one with water, one with syrup or shampoo). Ask the students, "What is the difference between these two liquids?"

4. Allow the children to explore the materials (without opening the bags) in small groups on the rug. Direct the students to turn the bags so that the liquids flow. Have them gently squeeze the bags and feel the difference between the materials. Record the student observations on the States of Matter Class Chart.

5. Direct the students to illustrate and write about two liquids in their *Matter* Science Journals (Observation 2, page 160). (Assessment)

Inquiry Skills—Observation, Compare and Contrast

Matter *(cont.)*

Lesson 6: Solids Demonstration/Activity

Materials (for each pair)

- container of play dough
- 3 wooden blocks

Directions

1. Provide each pair with a container of play dough. Allow the children to explore the material. Ask them, "Is this a liquid, solid, or gas? How do you know?"

2. Instruct the children to return the play dough back to the container and close the lid.

3. Provide the pairs with the wooden blocks. Have the children explore the materials. Ask them, "Is this a liquid, solid, or gas? How do you know? How is this solid different from the other solid (play dough)? Why are they both solids? What happens if we leave play dough out of its container overnight?"

4. Direct the students to illustrate and write about two solids in their *Matter* Science Journals (Observation 3, page 161). (Assessment)

Inquiry Skills—Observation, Compare and Contrast

Lesson 7: Gases Demonstration/Activity

Materials

- hot plate
- tea kettle
- water
- wrapped drinking straws
- piece of crépe paper for each student (cut into 3' [91.4 cm] lengths)

Safety Note: Instruct the students on proper use of straws.

Directions

1. Use a hot plate to heat water in a tea kettle. Have the students observe from a safe distance when the water boils and the steam leaves the kettle. Ask the children, "Is this a liquid, solid, or gas? How do you know? Is it still water?" (*Yes, it is water vapor, a type of gas.*)

2. Give a straw to each child. Instruct the children to carefully tear off only one end of the straw wrapper. Direct them to place the exposed end of the straw in their mouths and blow. Ask the students, "What happened?" (*The paper wrapper should have blown off.*) "Why did that happen?" (*Because the air blew it off!*)

3. Have the students work in pairs to gently blow air on each other's hands with a straw. Ask them, "What are you blowing on each other's hands?" (*air, a gas*)

4. Help each child tape a length of crépe paper onto the end of his or her straw. Take the children outside to observe wind conditions. Have the children wave their streamers in the air to observe the effect this has on the crépe paper. Ask the children, "Is this a liquid, solid, or gas? How do you know?"

5. Instruct the students to illustrate and write about two gases in their *Matter* Science Journals (Observation 4, page 162). (Assessment)

Inquiry Skills—Observation, Classifying, Compare and Contrast

Matter *(cont.)*

Lesson 8: Changes in States of Matter Demonstration

Materials

- water
- ice cube tray
- 3 plastic plates
- 8 oz. (237 mL) glass
- freezer

Directions

1. While the children are watching, pour water into an ice cube tray. Ask another adult to take the tray and place it in a freezer. At the end of the day, remove the tray and show it to the children. Ask the children, "What changes happened?" (*The water froze. It became a solid.*)

2. Place the ice cubes on plastic plates in various locations (outdoors in sunlight, outdoors in shade, in the classroom) and have the students observe changes over time.

3. Place a large glass of ice water on a table in the classroom where all children can see it while they are doing independent work. Point out that there is water inside the glass, but none on the outside of it. After a sufficient amount of time, observe the glass. Ask the students, "What has happened to the outside of the glass?" (*Water has collected on it.*) "Where did the water come from?" (*the air*) "Is the water liquid, solid, or gas?" (*liquid*) "Was the water in the air liquid, solid, or gas?" (*gas*)

4. Have the students write about the changes in states of matter they observed in their *Matter Science Journals* (Observations 5–7, page 163). (Assessment)

Inquiry Skill—Observation

Lesson 9: Mixing Liquids Investigation 1: Fun Goo (Liquid to Solid)

Materials (for each small group)

- quart-sized resealable plastic bag
- plastic spoon
- permanent marker
- Fun Goo Rebus Sheet (page 137)
- Fun Goo Group Response Sheet (page 144)

Ingredients

- 4 spoonfuls white glue
- 1 drop food coloring
- 2 spoonfuls Mystery Liquid #1 (see Unit Preparation, page 131, #6)

Step 1—Predict

1. Make a prediction about what will happen when you mix the ingredients together.
2. Have the Recorder write the group's response on the Fun Goo Group Response Sheet (page 144).

Step 2—Mix Liquids

1. Use the marker to write *Mixture #1* on the plastic bag.
2. Pour the glue into the bag.
3. Add 1 drop of food coloring to the glue.
4. Seal the bag and mix by gently kneading.
5. Add 2 spoonfuls of the mystery solution to the glue.
6. Seal the bag and knead to mix.

Step 3—Compare

1. Compare your prediction to what actually happened.
2. Have the Recorder write the group's response on the Fun Goo Group Response Sheet (page 144).

Inquiry Skills—Observation, Predicting, Compare and Contrast

Fun Goo Rebus Sheet

Step 1—Predict

Step 2—Mix Liquids

1.

2.

3.

4.

5.

6.

Step 3—Compare

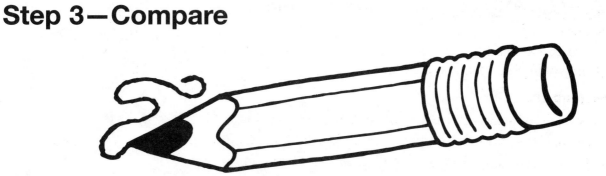

Matter *(cont.)*

Lesson 9: (cont.)
Mixing Liquids Investigation 2: Great Goo (Liquid to Solid)

Materials (for each small group)

- quart-sized resealable plastic bag
- plastic spoon
- permanent marker
- Great Goo Rebus Sheet (page 139)
- Great Goo Group Response Sheet (page 145)

Ingredients

- 4 spoonfuls white glue
- 1 drop food coloring
- 2 spoonfuls Mystery Liquid #2 (see Unit Preparation, page 131, #6)

Step 1—Predict

1. Make a prediction about what will happen when you mix the ingredients together.
2. Have the Recorder write the group's response on the Great Goo Group Response sheet (page 145).

Step 2—Mix Liquids

1. Use the marker to write *Mixture #2* on the plastic bag.
2. Pour the glue into the bag.
3. Add 1 drop of food coloring to the glue.
4. Seal the bag and mix by gently kneading.
5. Add 2 spoonfuls of the mystery solution to the glue.
6. Seal the bag and knead to mix.

Step 3—Compare

1. Compare your prediction to what actually happened.
2. Have the Recorder write the group's response on the Great Goo Group Response sheet (page 145).

Inquiry Skills—Observation, Predicting, Compare and Contrast

Great Goo Rebus Sheet

Step 1—Predict

Step 2—Mix Liquids

1.

2.

3.

4.

5.

6.

Step 3—Compare

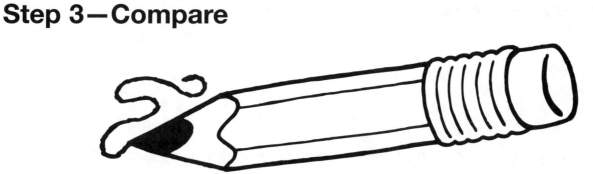

Matter *(cont.)*

Lesson 9: (cont.)
Mixing Liquids Investigation 3: Gel Goo (Liquid to Solid)

Materials (for each small group)

- quart-sized resealable plastic bag
- plastic spoon
- permanent marker
- Gel Goo Rebus Sheet (page 141)
- Gel Goo Group Response Sheet (page 146)

Ingredients

- 4 spoonfuls blue gel glue
- 1 drop food coloring
- 2 spoonfuls Mystery Liquid #1 (see Unit Preparation, page 131, #6)

Step 1—Predict

1. Make a prediction about what will happen when you mix the ingredients together.

2. Have the Recorder write the group's response on the Gel Goo Group Response sheet (page 146).

Step 2—Mix Liquids

1. Use the marker to write *Mixture #3* on the plastic bag.

2. Pour the glue into the bag.

3. Add 1 drop of food coloring to the glue.

4. Seal the bag and mix by gently kneading.

5. Add 2 spoonfuls of the mystery solution to the glue.

6. Seal the bag and knead to mix.

Step 3—Compare

1. Compare your prediction to what actually happened.

2. Have the Recorder write the group's response on the Gel Goo Group Response sheet (page 146).

Inquiry Skills—Observation, Predicting, Compare and Contrast

Note: Usually blue gel glue creates a clear or translucent goo rather then an opaque goo.

Gel Goo Rebus Sheet

Step 1—Predict

Step 2—Mix Liquids

1.

2.

3.

4.

5.

6.

Step 3—Compare

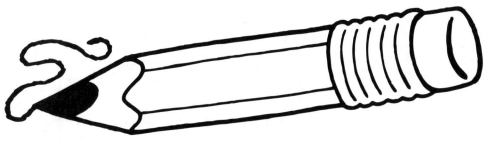

Matter *(cont.)*

Lesson 10: Mixing Liquid and Solid Investigation: Oobleck (Liquid to Solid)

Materials (for each small group)

- quart-sized resealable plastic bag
- plastic spoon
- Oobleck Rebus Sheet (page 143)
- Oobleck Group Response Sheet (page 147)

Ingredients

- 2 cups cornstarch
- 3 drops food coloring
- water, add slowly

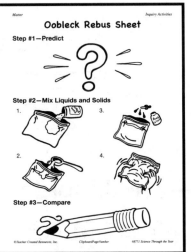

Step 1—Predict

1. Make a prediction about what will happen when you mix the ingredients together.

2. Have the Recorder write the group's response on the Oobleck Group Response Sheet (page 147).

Step 2—Mix Liquid and Solid

1. Carefully pour cornstarch into bag.

2. Slowly add water one spoonful at a time to the bag. When the cornstarch is no longer dry and flows, stop adding water.

3. Add 3 drops of food coloring to the mixture.

4. Seal the bag and mix by gently kneading.

Step 3—Compare

Compare your prediction to what actually happened.

1. Have the Recorder write the group's response on the Oobleck Group Response Sheet (page 147).

Inquiry Skills—Observation, Predicting, Compare and Contrast

Oobleck Rebus Sheet

Step 1—Predict

Step 2—Mix Liquids and Solids

1.

2.

3.

4.

Step 3—Compare

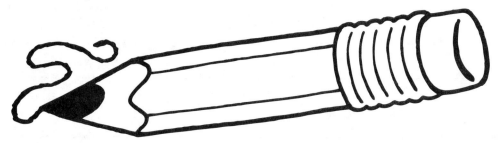

Mixing Liquids
Investigation 1: Fun Goo

Group Response Sheet

Predict: What will happen when you mix the liquids together?

_____

_____

_____

_____

_____

_____

_____

Compare: What actually happened?

_____

_____

_____

_____

_____

_____

_____

Mixing Liquids
Investigation 2: Great Goo

Group Response Sheet

Predict: What will happen when you mix the liquids together?

_____

_____

_____

_____

_____

_____

_____

Compare: What actually happened?

_____

_____

_____

_____

_____

_____

_____

Mixing Liquids
Investigation 3: Gel Goo

Group Response Sheet

Predict: What will happen when you mix the liquids together?

_____

_____

_____

_____

_____

_____

_____

Compare: What actually happened?

_____

_____

_____

_____

_____

_____

_____

Mixing Liquids and Solid Investigation: Oobleck

Group Response Sheet

Predict: What will happen when you mix the liquids and solid together?

_____

_____

_____

_____

_____

_____

Describe: Is the mixture a liquid or a solid? How do you know?

_____

_____

_____

Compare: What actually happened?

_____

_____

_____

_____

Matter *(cont.)*

Lesson 11: Literature Connection

1. Introduce the story *Bartholomew and the Oobleck* by Dr. Seuss to the class. Talk about the front cover of the book, the author, and illustrator.

2. Do a "picture walk" first. Show the children each page of the book without reading the text and ask them to predict what will happen at various points in the story.

3. Read the book to the class.

4. Ask the children to share what they liked best about the story.

5. Have the children help you complete a Plot Analysis Chart (see below). (Whole Class Assessment)

Sample Plot Analysis Chart: *Bartholomew and the Oobleck*

Beginning	Middle	End
The king is bored with the weather.	The royal magicians make a new type of weather—the Oobleck! But the Oobleck ruins everything.	Bartholomew gets the king to say he is sorry for the mess, and the Oobleck disappears.

Lesson 12: *States of Matter* Minibook and Big Book

1. Distribute a *States of Matter* Minibook (pages 164–166) to each child.

2. Read the book as a class. Discuss the illustrations. Point out key vocabulary (*liquid, solid, gas*). Review the *Matter* Word Cards (page 170).

3. Explain what the children are to do at each center.

   Center 1: Read the *States of Matter* Minibook in small groups with the teacher. Then independently read the *States of Matter* Minibook several times to develop fluency.

   Center 2: Color the pictures in the *States of Matter* Minibook.

   Center 3: Use a yellow crayon to highlight key vocabulary: *liquid, solid, gas.* Using a red crayon, circle the high frequency word *a*. Draw a box around the word *matter* each time you see it. (Assessment)

   Center 4: Listen to a taped version of *Bartholomew and the Oobleck* or the *States of Matter* Minibook.

4. At the end of the unit, each child will take home the *States of Matter* Minibook and "read" it to his or her family.

5. Create an additional class book by enlarging the *States of Matter* Minibook. Students can take turns coloring the pages. Place the completed *States of Matter* Big Book in the classroom library.

Matter (cont.)

Lesson 13: *Changes in States of Matter Minibook and Big Book*

1. Distribute a *Changes in States of Matter* Minibook (pages 167–169) to each child.

2. Read the book as a class. Discuss the illustrations. Point out key vocabulary (*heating, mixing, cooling*).

3. Explain what the children are to do at each center.

   Center 1: Read the *Changes in States of Matter* Minibook in small groups with the teacher.

   Center 2: Color the pictures in the *Changes in States of Matter* Minibook.

   Center 3: Use a yellow crayon to highlight key vocabulary: *heating, mixing, cooling*. Using a red crayon, circle the high frequency word *can*. Draw a box around the word *change* each time you see it. (Assessment)

   Center 4: Listen to a taped version of Solids, Liquids, and Gases by Louise Osborne or the Changes in States of Matter Minibook.

4. At the end of the unit, each child will take home the *Changes in States of Matter* Minibook and "read" it to his or her family.

5. Create an additional class book by enlarging the *Changes in States of Matter* Minibook. Students can take turns coloring the pages. Place the completed *Changes in States of Matter* Big Book in the classroom library.

Matter *(cont.)*

Culminating Activity: Making Ice Cream

Materials (for each pair)

- quart-sized resealable plastic bag
- gallon-sized resealable plastic bag
- plastic spoon
- dish towel
- Making Ice Cream Rebus Sheet (page 151)

Ingredients

- 2 oz. (59 mL) milk
- 1 teaspoon (5 mL) sugar
- ¼ teaspoon (1 mL) vanilla extract
- 4 teaspoons (20 mL) rock salt
- 1 cup (237 mL) ice

Directions

1. Add ingredients to the small bag. Seal the bag, allowing air to escape.
2. Carefully knead the bag to mix the ingredients.
3. Fill the large bag with ice.
4. Add rock salt to the bag.
5. Put the small bag into the large bag (still sealed).
6. Seal the large bag.
7. Place the large bag on the dish towel. Wrap the towel around bag.
8. Carefully knead the bag, shaking it until the milk freezes into ice cream.

* **Inquiry Skill—Observation**

Extension: Matter at Home

As an extension, parents can help children do baking and cooking at home to see changes in states of matter (see Recipe Cards and parent letter, pages 128–130).

- Chocolate Chip Cookies (liquid and solid ingredients mix to form a solid, heat further solidifies the matter)
- Juice Pops (liquid to solid)
- Frozen Water Bottle (liquid to solid)
- Gelatin (mix liquid and solid ingredients, heat mixture, and cool to solid)

Assessment Pages: Demonstrating Content Knowledge and Vocabulary

- **Sequencing Changes** (pages 152–154)—Sequence stories that show changes in states of matter.
- **What's the Matter?** (page 155)—Match each *States of Matter* word (*liquid, solid, gas*) to an appropriate example of matter.
- **Changes, Changes, Changes!** (page 156)—Write the change that is taking place in each picture. Use a *Changes in States of Matter* word (*heating, mixing, cooling*).
- **States of Matter** (page 171)—Cut and paste each *States of Matter* word (*liquid, solid, gas*) under a corresponding example.
- **Changes in States of Matter** (page 172)—Match each *Changes in States of Matter* word (*heating, mixing, freezing*) to the corresponding picture.

Making Ice Cream Rebus Sheet

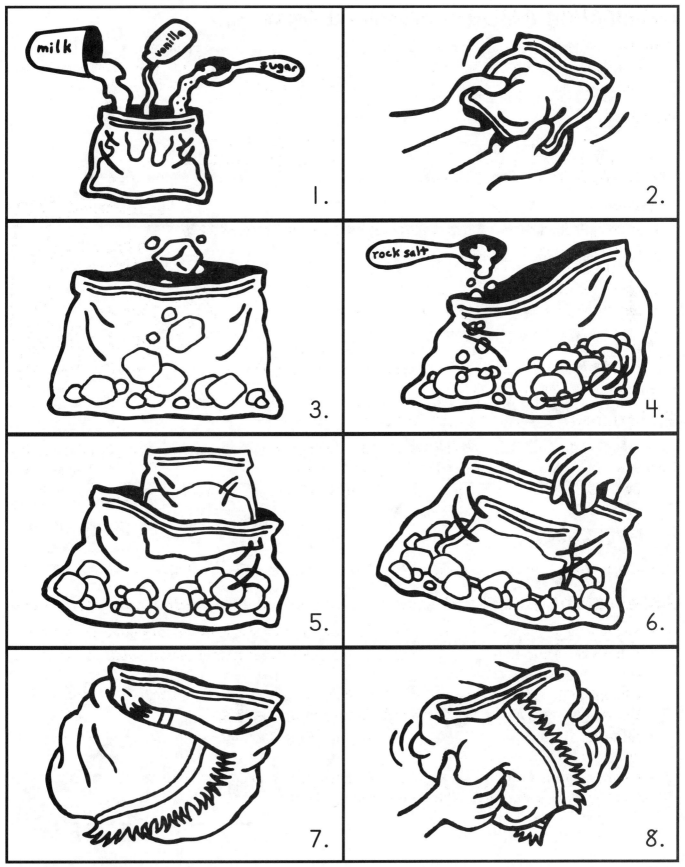

Matter *(cont.)*

Culminating Assessment: Sequencing Changes

Objective: Students will show the correct sequence for stories that show changes in states of matter.

Materials (for each child)

- Changes Story 1 and Changes Story 2 (pages 153–154)
- scissors
- crayons
- 11" x 17" (11.4 x 61 cm) construction paper, any color (see Unit Preparation, page 131)

Directions

1. Give each child a copy of one of the Changes Stories (pages 153–154).
2. Have the children color each picture on their pages.
3. Have the children carefully cut out each sequence card.
4. Direct the children to arrange the cards to create a story.
5. Instruct the children to glue the cards in the correct order on the construction paper.
6. Discuss the changes in matter in his or her story configuration.

Suggested Sequence (page 153)

Suggested Sequence (page 154)

Changes Story 1

One winter's day there was the first snow storm of the year. Lots and lots of snow fell.

At the end of the winter, the pond melted. It was spring. The children went fishing at the pond.

Some boys and girls made a snow boy. The weather was very cold, so the snow boy didn't melt.

Soon the weather became very warm. The children went swimming in the pond. It was summer.

Changes Story 2

One day, the sunlight began to melt the snow boy. Slowly, he melted into a puddle of water.

In the fall, the leaves began to fall off the trees and the pond was too cold to swim in.

The next day, the weather got cold again and the water froze into a huge pond of ice.

The weather grew colder, and soon the first snow of winter fell! The children made a snow girl!

What's the Matter?

Directions: Draw a line from each word to an example of that state of matter.

liquid

liquid

solid

solid

gas

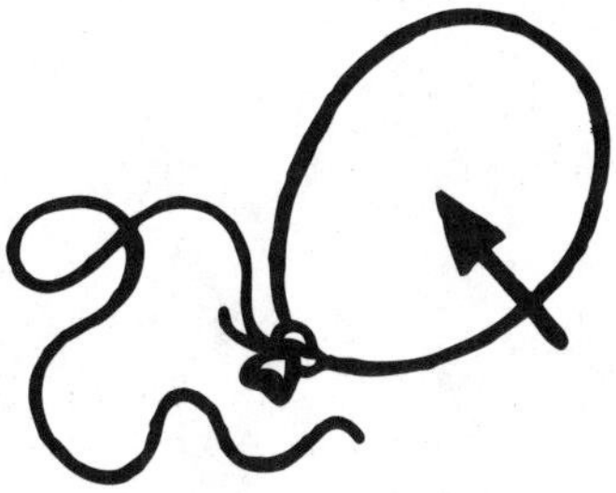

gas

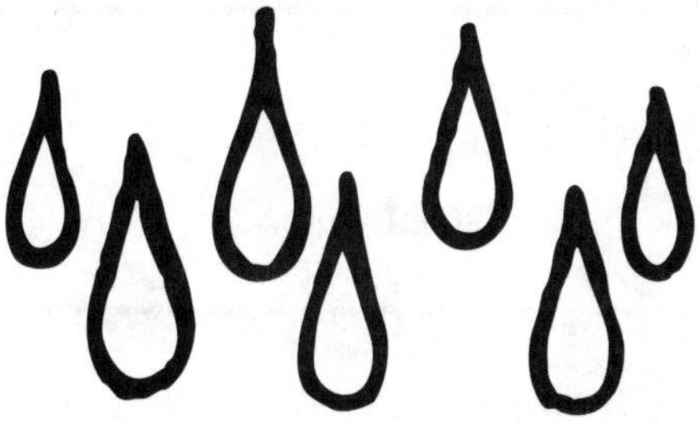

Changes, Changes, Changes!

Directions: Use the words in the Word Box. Write each change that is happening.

_____ _____

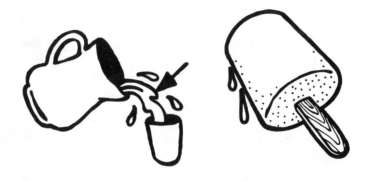

_____

Word Box

heating   freezing   mixing

Science Journal

MATTER

By:_____

 # Science Journal

Observation 1: Playing with Matter

1. Draw a picture of one example of matter from today's lesson.

2. Label your picture.

Describe: Write two sentences about your example of matter.

_____

_____

_____

Science Journal

States of Matter Chart

1. Add the words from the class chart to the top row below.

2. Draw an example of each state of matter in the bottom row.

3. Label your pictures.

Liquids	Solids	Gases
_____ _____ _____ _____ _____	_____ _____ _____ _____ _____	_____ _____ _____ _____ _____

Science Journal

Observation 2: Liquids

1. Draw a picture of a liquid.

2. Draw a picture of another liquid.

Liquid 1   **Liquid 2**

Compare and Contrast

What is the same about the liquids?

_____

_____

What is the difference between the liquids?

_____

_____

Science Journal

Observation 3: Solids

1. Draw a picture of a solid.

2. Draw a picture of another solid.

Solid 1   **Solid 2**

Compare and Contrast

What is the same about the solids?

_____

_____

What is the difference between the solids?

_____

_____

Science Journal

Observation 4: Gases

1. Draw a picture of a gas.

2. Draw a picture of another gas.

Gas 1   **Gas 2**

Compare and Contrast

What is the same about the gases?

_____

_____

What is the difference between the gases?

_____

_____

Science Journal

Observations 5–7: Changes in States of Matter

1. Draw a picture of each change you observed.

2. Label your pictures.

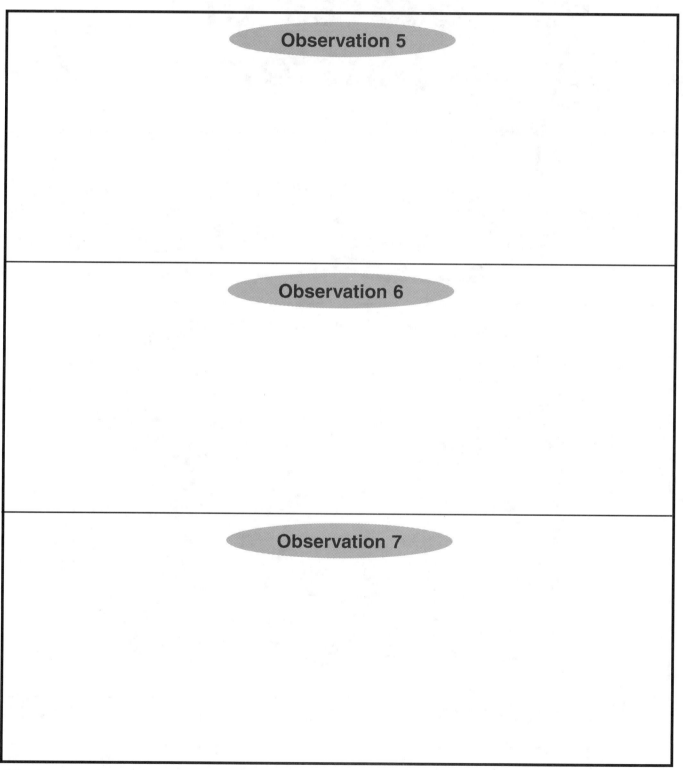

Observation 5

Observation 6

Observation 7

States of Matter

By: _____

Matter comes in different forms.

①

Matter can be a *liquid*.   **2**

Matter can be a *solid*.   **3**

Matter can be a *gas*. **4**

Everything on Earth is made of matter. **5**

Changes in States of Matter

By: _____

Matter can change states.

1

Heating matter can change it from a solid to a liquid.

2

Heating matter can change it from a liquid to a gas.

3

Sometimes mixing matter can change it from a liquid to a solid.

❹

Freezing matter can change it from a liquid to a solid.

❺

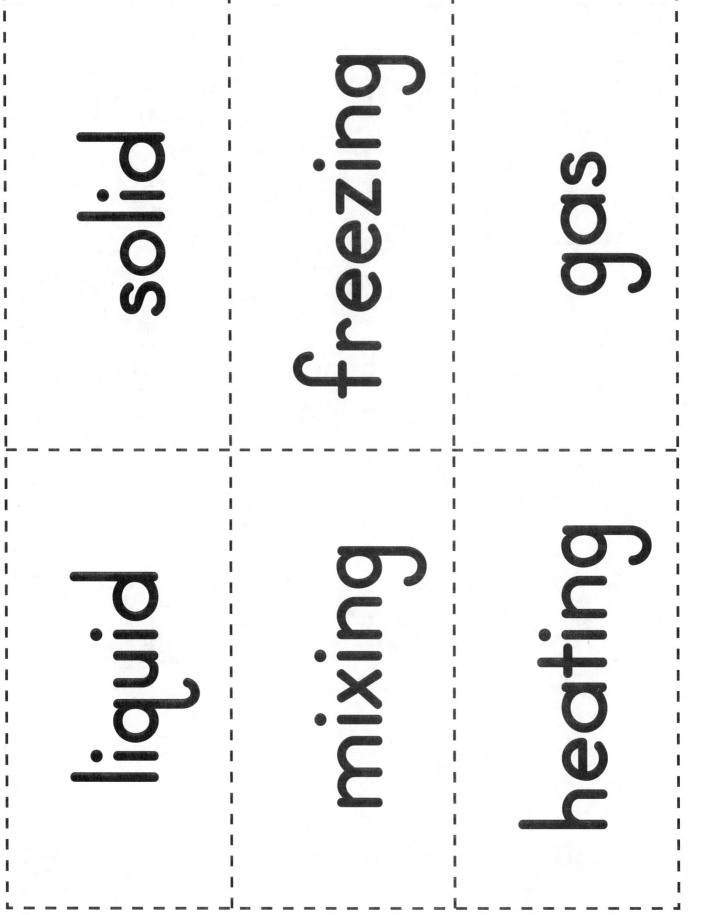

solid

freezing

gas

liquid

mixing

heating

States of Matter

Directions: Cut out the word cards. Glue each card under the correct picture.

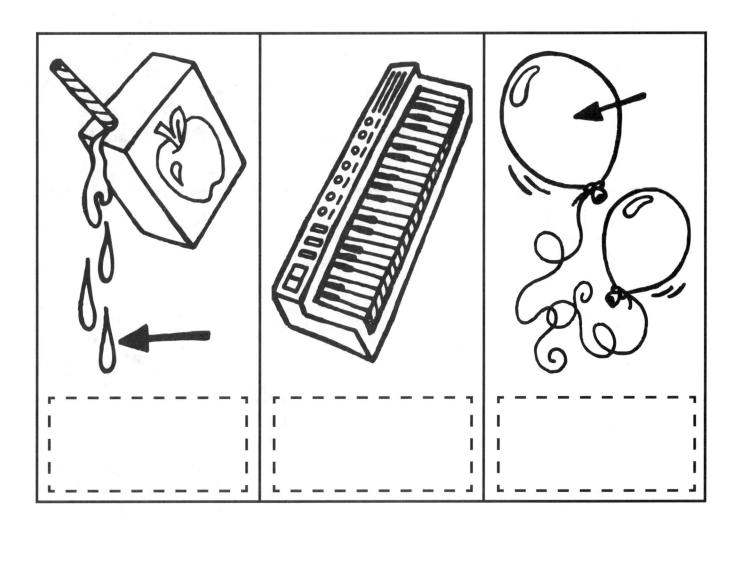

Changes in States of Matter

Directions: Draw a line to match each word to the correct picture.

heating

freezing

mixing

Germs

The illustrations below depict microscopic views of germs.
What differences do you see?

bacteria   viruses

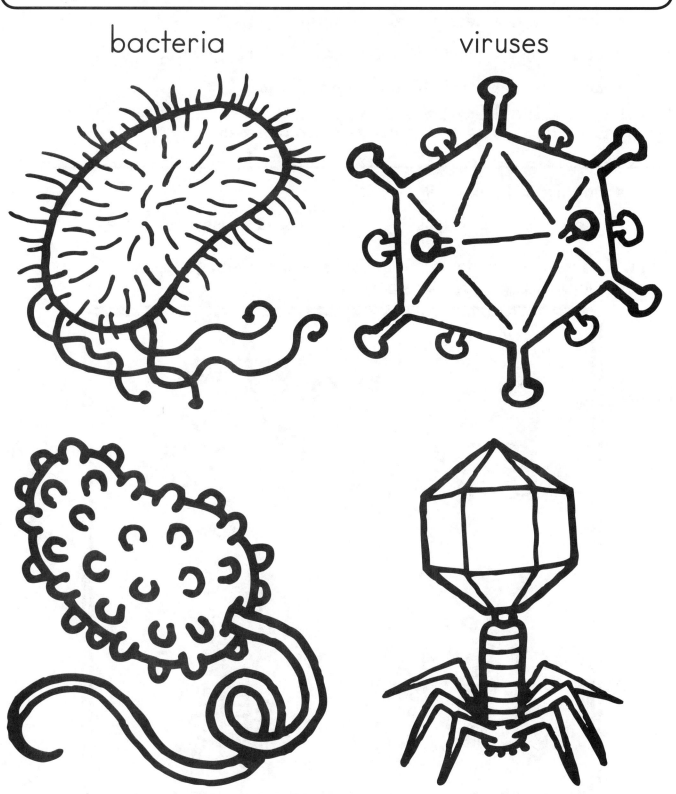

Germs Vocabulary

Germs—bacteria and viruses

- **Bacteria**—tiny germs that live in the air and on every surface

    - most are harmless and even helpful

    - some cause common infections (example: strep throat)

    - antibiotics help rid the body of bacteria

- **Viruses**—tiny germs that cause common diseases
  (example: flu [causes fever, body aches, coughing])

    - antibiotics do not kill viruses

    - there are vaccinations for some diseases caused by viruses

Note: The illustrations below are enlargements of what germs (viruses and bacteria) look like under a microscope.

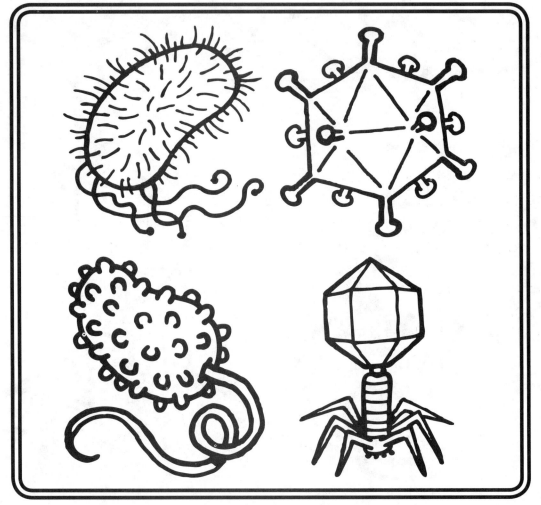

- -

Date _____

Dear Parent,

We will be starting a science unit on germs soon. There are a few items we need for the unit. Please provide the item circled below. If you are unable to provide the item, please let me know as soon as possible.

- box of quart-sized resealable plastic bags

- loaf of white bread (preferably with no preservatives)

- disposable gloves

Please send your item to school with your child by _____.
Thank you for your help with this unit!

Sincerely,

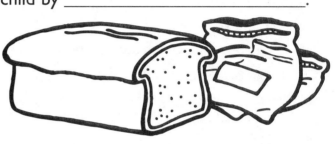

- -

Date _____

Dear Parent,

We have just started a science unit on germs! We will be learning ways to prevent disease and maintain good health. Please ask your child what kinds of germs there are and what he or she can do to stay healthy over the next few weeks.

At the end of the unit, students will read the *Germs* Minibook. Please ask your child to "read" his or her book to you and tell you what he or she learned about staying healthy.

Sincerely,

- -

Germs

 ## Background Information

Personal health and disease prevention are important areas of study for young children because elementary school children get colds and flu often. Disease prevention involves learning about germs. *Germs* are microorganisms carried by animals or through the air. Germs are bacteria or viruses that cause common illnesses such as colds and the stomach flu.

 ## Unit Preparation

1. Purchase the following items at a grocery or discount store or ask for donations (see parent letter on page 175):

    - box of quart-sized resealable plastic bags

    - loaf of white bread (preferably with no preservatives)

    - disposable gloves

    - kitchen tongs

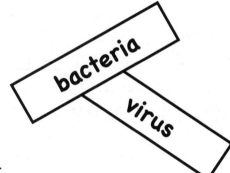

2. Copy and cut out the *Germs* Word Cards (page 188).

3. Copy and assemble the *Germs* Minibooks (pages 185–187).

4. Copy and assemble the *Germs* Science Journals (pages 182–184).

5. Reproduce the *Germs* Assessment Sheet (page 181).

6. Use chart or butcher paper to create the Germs Inquiry Chart for Lesson 1 (see page 177), as well as the charts used throughout the unit.

Literature Links

The Berenstain Bears Go to the Doctor by Stan and Jan Berenstain

Eyewitness Books: Epidemic by Brian Ward

Germs Are Not for Sharing by Elizabeth Verdick

Germs Make Me Sick! by Melvin Berger

Miss Bindergarten Stays Home from Kindergarten by Joseph Slate

What to Expect When You Go to the Doctor by Heidi Murkoff

Germs *(cont.)*

Lesson 1: Introducing the Germs Unit

Ask the children if they have heard of germs. Ask each student to turn to a neighbor and tell him or her what the student know about germs. Allow a few students to share their examples with the class. Tell them that they are starting a science unit about germs, but first they need to find out what they already know. Show the class the Germs Inquiry Chart (see below) and ask them to help you complete each section of the chart. (Whole Class Assessment)

Note: Color-code each column of the chart using bright markers such as red, blue, and green.

Germs Inquiry Chart

What Do You Know About Germs?	What Do You Want to Learn About Germs?	How Can We Find Out?

The inquiry chart serves as an assessment because it provides a good indication of what the children already know, what their misconceptions might be, and what they are wondering about. It is an excellent springboard for future lessons.

This sample chart gives an idea of the kinds of responses first grade children might give.

Sample Germs Inquiry Chart

What Do You Know About Germs?	What Do You Want to Learn About Germs?	How Can We Find Out?
You can't see germs. Germs can make you sick. You have to wash your hands. If someone who is sick drinks from a cup of water, and then you drink it, you can get sick.	What are germs? How small are germs? Why do germs come?	Read books. Go to the library. Check the Internet. Ask our parents. Ask the teacher.

Note: At the end of each unit, you can have the children help you complete a chart of "What We Learned" as a whole group culminating assessment.

Germs *(cont.)*

Lesson 2: Moldy Bread Bags

Materials (for each pair)

- 2 quart-sized resealable plastic bags (one labeled *dirty* and one labeled *clean*)
- 2 slices of white bread (preferably with no preservatives)
- 2 pairs of disposable plastic gloves and kitchen tongs

Note: This activity is best done after recess, before the children have
had a chance to wash their hands.

1. Distribute materials to the pairs. Give only one slice of bread at first.

2. Have the children gently rub their hands on one slice of the bread
   (both sides). They need to do this carefully so they do not tear the bread.

3. Have the children carefully insert that slice of bread into the bag
   marked *dirty* and seal it. Direct the children to set aside the bag.

4. Have the children wash their hands.

5. The teacher uses clean kitchen tongs to distribute a clean piece of bread to each pair. Have the
   children open the bag marked *clean*, the teacher places the bread in it, and the children seal the
   bag without touching the bread.

6. Have the children take out their *Germs* Science Journals and draw a picture of each bag and
   describe what they think will happen to each one after a week (Observation 1, page 183).
   (Assessment)

7. Direct the students to observe the bags daily for one week. At the end of a week, students
   record what happened to each bag in their *Germs* Science Journals (Observation 2, page 184).
   (Assessment)

8. After about one week, mold should grow on the bread in the bag marked *dirty* and there should be
   no mold or less mold on the bread in the bag marked *clean*. Lead a class discussion on why the
   dirty bread grew more mold. Talk about how the germs on the students' hands caused the mold
   to grow. Discuss with the children what mold is and why the germs on the dirty hands helped
   the mold grow. Explain that mold is not a plant or an animal, it is a fungus. The dirt and germs
   (bacteria) on the students' hands helped the mold grow.

Inquiry Skills—Observation, Predicting, Comparing and Contrasting

Lesson 3: Literature Connection

Note: This book contains a lot of information. You might choose to read the book in sections and
discuss it with the whole class as you go along. Or you could read it in small groups.

1. Introduce the story *Germs Make Me Sick!* by Melvin Berger to the class. Talk about the front
   cover of the book and the author and illustrator.

2. Do a "picture walk" first. Show the children each page of the book without reading the text and
   ask them to predict what will happen at various points in the story.

3. Read the book to the class.

4. Ask the children to share interesting facts they learned about germs.

5. As a class, discuss what the children learned about germs. Record their responses on chart paper.
   (Whole Class or Small Group Assessment)

Germs (cont.)

Lesson 4: Germs Picture Chart

1. Gather the children together on the rug near the *Germs* Picture Chart (see page 173) to teach content and vocabulary. Show the children the chart paper with the pre-drawn pencil diagram of different types of germs. (Be sure to write your teacher notes lightly in pencil next to each part of the diagram.)

2. Talk about germs in "chunks," providing key vocabulary (*bacteria, viruses*) and content (*bacteria cause infections such as the strep throat, viruses cause the common cold and the flu*). As you speak, trace over the corresponding parts with a permanent marker. Do not label the diagram yet.

   Teacher Note: Describe what you are doing as you draw. Mention the curving lines, larger and smaller shapes, etc. Your descriptions should help students feel more comfortable with their own Science Journal illustrations.

3. After tracing over the entire picture and providing the vocabulary and content, ask the children to tell what they learned about germs.

4. Write the children's responses on appropriate places on the chart (i.e., labeling types of germs).

5. Introduce the *Germs* Word Cards (page 188) and add them to a pocket chart.

6. Display the completed *Germs* Picture Chart in the classroom for a reference tool.

 Germs Home Link: Have the children complete the *Wash Your Hands* worksheet (page 189) for homework.

Lesson 5: Importance of Doctor Visits

1. Have a whole group discussion about the importance of regular doctor visits.

   Make sure you cover these important points: Doctor visits help you prevent disease and help you maintain good health.

2. Give the children the opportunity to discuss their experiences at the doctor.

3. Have the children brainstorm a class list of the important things doctors do for children at their checkups. Record their answers on the board (see sample responses below).

Samples of How Doctors Help Children at Checkups

- check vision
- check hearing
- monitor height and weight (growth)
- ask about diet and exercise
- check heart, lungs, ears, mouth
- ask if there are any health problems
- give vaccinations (shots)

Germs *(cont.)*

Lesson 6: How to Prevent Disease

Over the course of the unit, read books to the class that share ways to protect yourself and others from spreading disease. (See Literature Links on page 176.) At the end of the unit, have the children brainstorm a list of ways to prevent disease. Record student responses on the board (see sample responses below).

Inquiry Skills—Observation, Recording Data

Examples of How Children Can Help Prevent Disease

- cover mouth when you cough or sneeze (sneeze into sleeve—not hand)
- wash hands after you sneeze
- wash hands after using the restroom
- wash hands before eating
- do not share cups, forks, spoons, toothbrushes
- use a tissue when you sneeze or blow your nose

Lesson 7: *Germs* Minibook and Big Book

1. Distribute a *Germs* Minibook (pages 185–187) to each child.

2. Read the book as a class. Discuss the illustrations. Point out key vocabulary (*bacteria, viruses*). Review the *Germs* Word Cards (page 188).

3. Explain what the children are to do at each center.

    Center 1: Read the *Germs* Minibook with a partner. Take turns reading each page of the book.

    Center 2: Color the pictures in the *Germs* Minibook.

    Center 3: Use a yellow crayon to highlight key vocabulary: *bacteria, virus*. Using a red crayon, circle the high frequency word *our*. Draw a box around the word *germs* each time you see it. (Assessment)

    Center 4: Listen to a taped version of *Germs Make Me Sick!* or the *Germs* Minibook.

4. At the end of the unit, each child will take home the *Germs* Minibook and "read" it to his or her family.

5. Create an additional class book by enlarging the *Germs* Minibook. Students can take turns coloring the pages. Place the completed *Germs* Big Book in the classroom library.

Assessment Pages: Demonstrating Content Knowledge and Vocabulary

- **Stop the Germs!** (page 181)—Color pictures to show behaviors that will prevent disease.
- **Wash Your Hands** (page 189)—Choose activities before or after which one should wash one's hands.

Stop the Germs!

Directions: Color each picture that shows a way to *prevent* disease.

Science Journal

GERMS

By: _____

Science Journal

Observation 1: Moldy Bread Bags

1. Draw a picture of the two bread bags.

2. Label each picture.

Prediction: What do you think will happen to the bread?

_____

_____

_____

Science Journal

Observation 2: Moldy Bread Bags 2

1. Draw pictures of the two bread bags.

2. Label each picture.

Describe: What happened to the bread? Write two or three sentences.

_____

_____

_____

_____

Germs can be harmless or harmful.   ❶

Germs can cause colds. A-choo! ❷

Germs can cause the stomach flu. ❸

Washing our hands keeps us from getting sick.

4

Covering our coughs helps us keep others from getting sick.

5

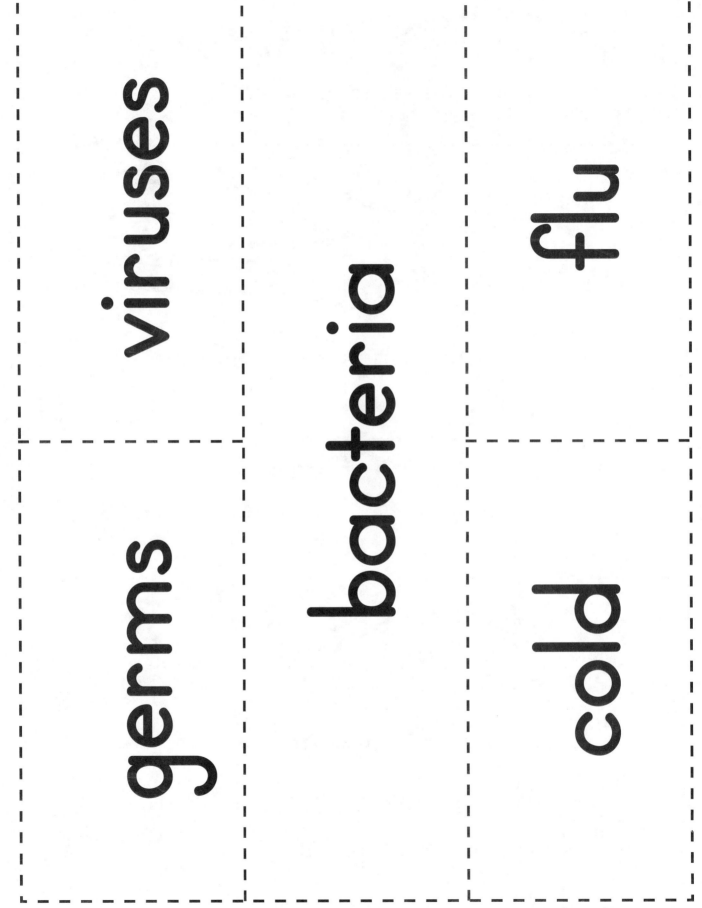

viruses

bacteria

flu

germs

cold

Wash Your Hands

Directions: Cut out each picture that shows when you should wash your hands. Glue the pictures in the boxes above.

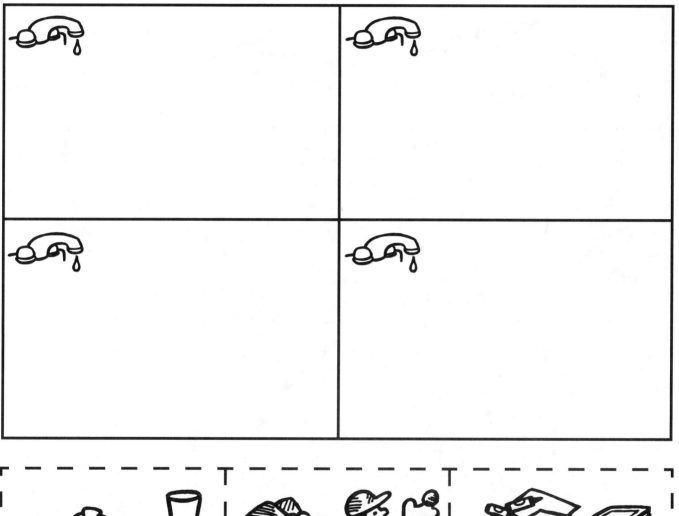

Parts of a Tooth

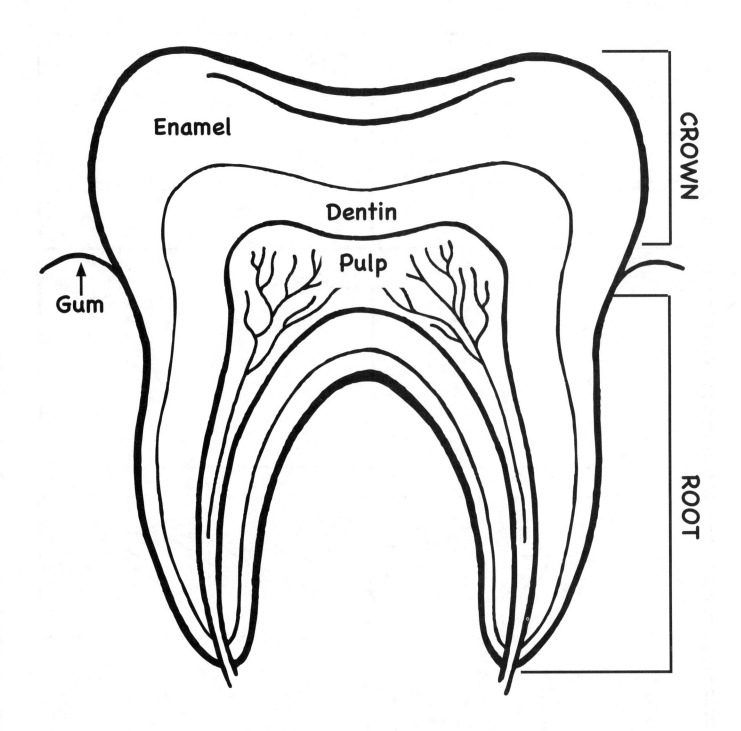

Parts of a Tooth Vocabulary

Crown—the part of the tooth you see when you smile

Dentin—yellow bone-like material, contains some nerves so you can feel if something is wrong with the tooth

Enamel—hard, white covering of the tooth

Pulp—soft tissue inside the tooth, has blood vessels so the tooth can receive nutrients and nerves so you can feel your tooth

Root—the part of the tooth below the gum line

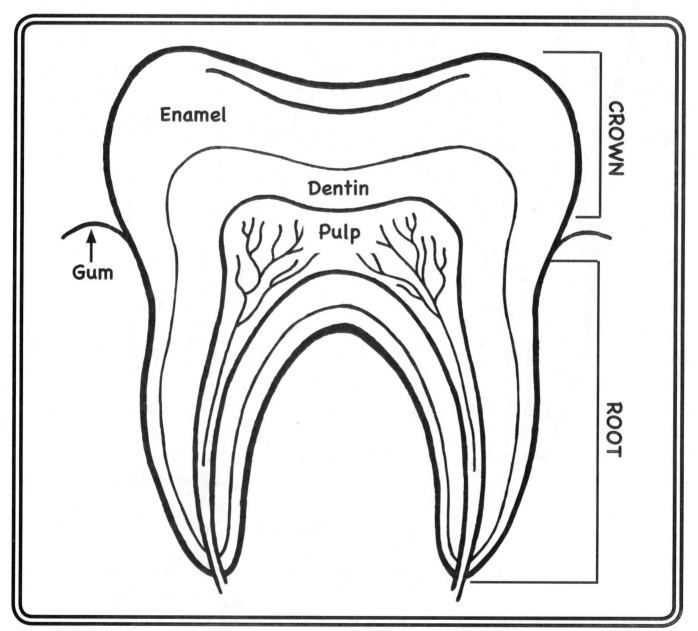

- -

Date _____

Dear Parent,

We will be starting a science unit on teeth soon. There are a few items we need for the unit. Please provide the item circled below. If you are unable to provide the item, please let me know as soon as possible.

- bag of marshmallows
- 4 apples
- bunch of celery
- bag of carrot sticks

- box of chocolate cookies
- bag of hard candy
- small plastic cups

Please send your item to school with your child by _____.

Thank you for your help with this unit!

Sincerely,

_____

- -

Date _____

Dear Parent,

We have just started a science unit on teeth! We will be learning ways to maintain healthy teeth. Please ask your child what he or she can do to take care of his or her teeth over the next few weeks.

At the end of the unit, students will read the *Healthy Teeth* Minibook. Please ask your child to "read" his or her book to you and tell you what he or she learned about maintaining healthy teeth.

Sincerely,

_____

- -

Teeth

Background Information

Personal health involves learning about how to take care of oneself, including proper care and cleaning of one's teeth and good hygiene. In this unit, the children will learn about the parts of a tooth and about baby and adult teeth.

Unit Preparation

1. Order educational materials from toothpaste companies before the school year begins.*

2. Purchase the following items at a grocery or discount store or ask for donations (see parent letter on page 192):

   - bag of marshmallows
   - box of raisins
   - 4 apples
   - bunch of celery
   - bag of carrot sticks
   - box of chocolate cookies
   - bag of hard candy
   - small plastic cups

3. Copy and cut out the *Teeth* Word Cards (page 207).

4. Copy and assemble the *Healthy Teeth* Minibooks (pages 204–206).

5. Reproduce the *Teeth* Assessment sheet (page 200).

6. Use chart or butcher paper to create the Teeth Inquiry Chart for Lesson 1 (see page 194), as well as the charts used throughout the unit.

---

*Check out *www.crest.com* for links to a classroom experiment using eggs, vinegar, and Crest® toothpaste, tips for healthy teeth, and free materials for first grade teachers. The kit includes a teacher's guide, toothpaste samples and toothbrushes, plaque disclosing tablets, a video, a CD, and information booklets for parents and children. Go to Colgate's website at *www.colgate. com* for downloadable instructional materials for primary grade teachers and online activities for children. Oral-B® also has a website at *www.oralb.com.* This site has printable handouts and lesson plans for teachers and interesting information such as the history of the toothbrush.

Literature Links

The Berenstain Bears Visit the Dentist by Stan and Jan Berenstain

How Many Teeth? by Paul Showers

Show Me Your Smile: A Visit to the Dentist by Christine Ricci

Taking Care of Your Teeth by Mary Elizabeth Salzman

What to Expect When You Go to the Dentist by Heidi Murkoff

Teeth (cont.)

Lesson 1: Introducing the Healthy Teeth Unit

Tell the children that they are going to start a unit on teeth. Ask each child to turn to a neighbor and tell him or her what the child knows about teeth. Allow a few students to share their examples with the class. Show the class the Teeth Inquiry Chart (see below) and ask them to help you complete each section of the chart.

Note: Color-code each column of the chart using bright markers such as red, blue, and green.

Teeth Inquiry Chart

What Do You Know About Teeth?	What Do You Want to Learn About Teeth?	How Can We Find Out?

The inquiry chart serves as an assessment because it provides a good indication of what the children already know, what their misconceptions might be, and what they are wondering about. It is an excellent springboard for future lessons.

This sample chart below gives an idea of the kinds of responses first grade children might give.

Sample Teeth Inquiry Chart

What Do You Know About Teeth?	What Do You Want to Learn About Teeth?	How Can We Find Out?
Sometimes they fall out. We chew our food with them.	Why do they fall out? How do they get cavities?	Go to the library. Ask a teacher. Ask a parent. Ask a dentist.

Note: At the end of each unit, you can have the children help you complete a chart of "What We Learned" as a whole group culminating assessment.

Teeth (cont.)

Lesson 2: Parts of a Tooth Picture Chart

1. Gather the children together on the rug near the *Parts of a Tooth* Picture Chart (see page 191) to teach content and vocabulary. Show the children the chart paper with the pre-drawn pencil diagram of a tooth. (Be sure to write your teacher notes lightly in pencil next to each part of the diagram.)

2. Talk about the parts of a tooth in "chunks," providing key vocabulary (*root, enamel, dentin, crown, pulp*) and content (*enamel is the outer coating of a tooth, the root connects the tooth to the gums*). As you speak, trace over the corresponding parts with a permanent marker. Do not label the diagram yet.

    Teacher Note: Describe what you are doing as you draw. Mention the curving lines, larger and smaller shapes, etc. Your descriptions should help students feel more comfortable with their own Science Journal illustrations.

3. After tracing over the entire picture and providing the vocabulary and content, ask the children to tell what they learned about teeth.

4. Introduce the *Teeth* Word Cards (page 207) and add them to a pocket chart.

5. Write the children's responses on appropriate places on the chart (i.e., labeling parts of a tooth).

6. Display the completed *Parts of a Tooth* Picture Chart in the classroom for a reference tool.

 Teeth Home Link: Have the children complete the *Parts of a Tooth* worksheet (page 208) for homework.

Lesson 3: Literature Connection

1. Introduce the story, *Show Me Your Smile: A Visit to the Dentist*, by Christine Ricci to the class. Talk about the front cover of the book and the author and illustrator.

2. Do a "picture walk" first. Show the children each page of the book without reading the text and ask them to predict what will happen at various points in the story.

3. Read the story to the class.

4. Ask the children to share what they liked best about the story.

5. As a class, have the children discuss what they learned about visiting the dentist.

6. Record student responses on chart paper (see sample responses below). (Whole Class Assessment)

Sample Responses: Visiting the Dentist

- It can be fun to visit the dentist!
- The dentist gives you a new toothbrush.
- If you don't have a cavity, you get to choose a sticker.
- My mouth feels really clean after I go for my checkup!
- The dentist helps me keep a healthy smile!

Teeth *(cont.)*

Lesson 4: Importance of Dental Visits

1. Have a whole group discussion about the importance of regular dental visits.

    Make sure you cover these important points: Dental visits help you prevent tooth decay and tooth loss and help you maintain a healthy smile.

2. Give the children the opportunity to discuss their experiences at the dentist.

3. Have the children brainstorm a class list of the important things dentists do for children at their checkups. Record student responses on the board (see sample responses below).

Samples Responses: How Dentists Help Children During Checkups

- take X-rays
- check for cavities
- clean teeth
- count teeth
- check baby teeth versus adult teeth
- check to see that teeth are growing in straight
- show you how to brush and floss
- give you a free toothbrush

Lesson 5: Class Tooth Chart

At the beginning of the school year, create a Class Tooth Chart (see below) so that you can record the number of teeth lost during the year. (It is best if the chart is divided into months.) Each time a child loses a tooth, attach a tooth cutout with the child's name on it and the date. At the end of each month, count the total number of teeth lost and make a separate graph. Explain that the teeth the children lose are baby teeth and the new teeth that replace them are adult teeth.

Inquiry Skills—Observation, Measurement

Sample Class Tooth Chart

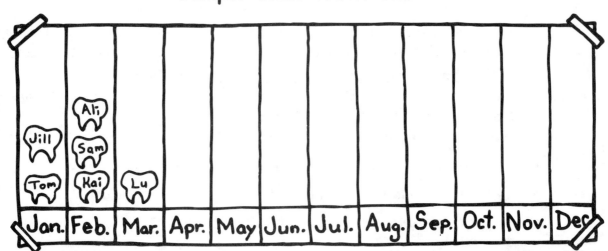

Teeth *(cont.)*

Lesson 6: Counting Our Teeth

Have the children work in pairs to help each other count their teeth. Provide the children with hand mirrors so that they can look into their mouths and count their teeth. Have the students assist you in recording the number of teeth each child has on a class graph.

Inquiry Skills—Observation, Measurement, Recording Data

Lesson 7: How to Take Care of Your Teeth

Lead a class discussion about ways to take care of your teeth.

Make sure you cover these important points:

- brushing twicc a day with fluoride toothpaste
- flossing every day
- visiting the dentist every six months
- eating healthful foods
- getting enough fluoride (taking a daily fluoride supplement or drinking fluoridated tap water)

Then invite a guest speaker, such as a local dentist or dental hygienist, to speak to the class about ways to take care of teeth. The presentation should include proper ways to brush and floss teeth.

Lesson 8: *Healthy Teeth* Minibook and Big Book

1. Distribute a *Healthy Teeth* Minibook (pages 204–206) to each child.

2. Read the book as a class. Discuss the illustrations. Point out key vocabulary (*healthy, brush, floss*). Review the *Teeth* Word Cards (page 207).

3. Explain what the children are to do at each center.

   Center 1: Read the *Healthy Teeth* Minibook in small groups with the teacher. Then independently read the *Healthy Teeth* Minibook several times to develop fluency.

   Center 2: Color the pictures in the *Healthy Teeth* Minibook.

   Center 3: Use a yellow crayon to highlight key vocabulary: *brush, floss*. Using a red crayon, circle the high frequency word *our*. Draw a box around the word *teeth* each time you see it. (Assessment)

   Center 4: Listen to a taped version of *The Berenstain Bears Visit the Dentist* by Stan and Jen Berenstein or the *Healthy Teeth* Minibook.

4. At the end of the unit, each child will take home the *Healthy Teeth* Minibook and "read" it to his or her family.

5. Create an additional class book by enlarging the *Healthy Teeth* Minibook. Students can take turns coloring the pages. Place the completed *Healthy Teeth* Big Book in the classroom library.

Teeth (cont.)

Culminating Activity: Healthy Teeth

Safety Note: Check for food allergies before this activity.

Materials (for each group of four students)

- paper plate
- 4 marshmallows
- 4 apple slices
- 4 celery sticks
- 4 carrot sticks
- 4 hard candies
- 4 chocolate cookies
- 4 small plastic cups (filled with water)

Directions

1. Distribute the materials to Materials Managers.

2. Explain to the students that they need to sample the items as a group (do not go ahead) and finish each part of their journal entry before they go on to the next item.

3. Have each student take one of the food items. Instruct him or her to carefully look at it. Direct the students to illustrate the items in their *Teeth* Science Journals (Observation 1, page 202).

4. Then direct the students to eat the items. After the students have eaten and carefully chewed an item, they record how their teeth feel in their *Teeth* Science Journals (Observation 2, page 203). (Assessment) Possible response: *Marshmallows stick to the teeth. Apples leave the teeth feeling clean.* After each food item is eaten, the students should take a drink of water to cleanse their mouths.

Teeth (cont.)

Culminating Activity: Healthy Teeth

Directions *(cont.)*

5. After each group has finished, hold a class discussion of which foods are healthy for the teeth and which are bad for the teeth. Record results on a class chart (see sample responses below). Talk about the importance of brushing one's teeth after sticky or sweet snacks and treats. Talk about the importance of choosing healthful snacks, such as fruits and vegetables, instead of sweets.

Inquiry Skills—Observation, Compare and Contrast, Recording Data

Sample Responses: Healthful Snack or Treat?

Healthy Snacks	Treats
• apples • carrots • celery	• cookies • marshmallows • candy

Assessment Pages: Demonstrating Content Knowledge and Vocabulary

- **Healthy Teeth** (page 200)—Color each picture that shows a way to keep your teeth healthy.

- **Parts of a Tooth** (page 208)—Color and label the parts of a tooth.

Healthy Teeth

Directions: Color each picture that shows a way to keep your teeth healthy.

Science Journal

TEETH

By:_____

Science Journal

Observation 1: Healthful Foods

1. Draw pictures of healthful foods you like to eat.
2. Label each picture.

Science Journal

Observation 2: Healthful Foods II

1. Eat each food and then fill in the chart.

How did your teeth feel after you ate each food?

Compare and Contrast

Did all the foods make your teeth feel the same or different?

_____

_____

_____

Healthy Teeth

By: _____

Teeth are important.

1

Teeth help us chew our food. ❷

We need to brush our teeth twice a day. ❸

We need to floss our teeth once a day.   **4**

If we take care of our teeth,
they will be healthy.   **5**

cavity

root

dentin

crown

enamel

pulp

Parts of a Tooth

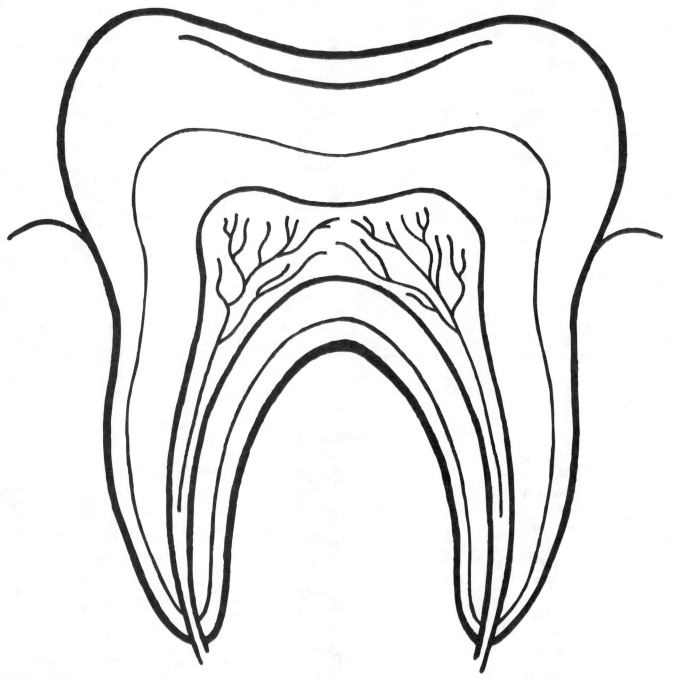

Directions: Color and label the parts of the tooth.

1. Color the *enamel* with a white crayon.

2. Color the *dentin* with a yellow crayon.

3. Color the *pulp* with a red crayon.

4. Label the *crown* and the *root*.